Sexuality Explained

Sexuality Explained

a guide for parents and children

Louise Kirk

Illustrated by Jessie Gillick

GRACEWING

First published in England in 2013
by
Gracewing
2 Southern Avenue
Leominster
Herefordshire HR6 0QF
United Kingdom
www.gracewing.co.uk

No part of this publication may be reproduced, stored in a retrieval system, or transmitted in any form or by any means, electronic, mechanical, photocopying, recording or otherwise, without the written permission of the publisher.

The right of Louise Kirk to be identified as the author of this work has been asserted in accordance with the Copyright, Designs and Patents Act 1988.

© 2013 Louise Kirk

ISBN 978 085244 808 3

Typeset by Gracewing

Cover design by Bernardita Peña Hurtado

Back-cover picture courtesy of Simon Caldwell, St Gabriel News and Media

Contents

Contents   5

Introduction   7

Principally for girls

1. Josie and the Cakes   13

General introduction to the female menstrual cycle: designed for pre-puberty.
9-11 years

2. Josie Lends a Hand   23

Review of the menstrual cycle in more practical detail for a girl approaching puberty.
10-12 years

3. Josie's Next Lesson   35

Explanation of the mucus cycle and its purpose.
11-13 years

4. A Privileged Role   47

Overview of male fertility and the process of conception.
12-14 years

5. It Helps to Know   61

Dress and teenage relationships; explanation of the chemistry of sexual bonding.
13-15 years

Principally for boys

6. Michael and Dad Go Fishing   75

General introduction to male fertility.
10-11 years

7. Michael Shows Off his Knowledge   87

Overview of female fertility and the process of conception.
12-14 years

8. Changed for Life 103
Sexual bonding, self-mastery and an overview of brain chemistry; an initial response to same-sex attraction.
13-15 years

FOR BOTH SEXES

9. Dad, Is Sex Dangerous? 117
Explanation of sexually transmitted diseases; the complementarity between the sexes.
13-15 years

10. Forgotten Truths 131
Overview of contraception and how it works; understanding how fertility can be managed naturally.
13-16 years

Endnotes 148

Whole page diagrams 151

Introduction

"Dad, what's sex?"

Sex education comes in many forms and, like it or not, you, as a parent, are going to be in the hot seat. Much of what you teach will be passed on subconsciously, but there will come a moment when you will want to open up specifically on the subject, so that you remain your children's first mentor. To be so, you will also want to know enough to be more convincing than anybody else.

Many parents find it awkward to discuss sex with their children, partly because their relationship has included few sexual references until then, and partly because of a natural reluctance to discuss something so private. This is entirely normal, which is why this Guide has been written—to help parents with the information they need, and to demonstrate some of the ways in which it can be imparted. There is a vast variety of 'right' ways of doing this, for which reason the Guide is not intended to be a 'how to do it' course.

It is instead designed to accompany parents and guardians throughout the years during which their children are growing into maturity. The Guide teaches by example and also by encompassing a much wider body of recent knowledge than most people have learnt. Ten chapters build upon each other to inspire children with wonder and to convince them that their own persons are worthy of the greatest respect. The conversational style accompanied by informal pictures and diagrams takes away embarrassment and makes the biology easy to absorb and pass on.

There are many ways to use the Guide, as outlined below, but, however you use it, your children will know that you care enough for them to teach them about sex yourself. This in itself will allow them to mature at their own rate, and to hold their own with their peers. It will also increase their trust of your judgment and deepen the lifelong bonds of your friendship.

Using the Guide

Getting started

The ten chapters which make up the Guide start from biology to demonstrate how body, mind, and spirit are integrated within human sexuality. Although the different chapters are designed to be absorbed by children over a period of years, it is recommended that you begin by reading the whole Guide. This will help you to decide when it may be appropriate to tackle what. The age range set against the different sections is for guidance only.

The best way to catch a child's attention is to deal with a topic when it is most relevant. This means noticing their and their friends' advance into puberty and being ready to respond to their questions. It also means knowing when sexual topics may be raised at school, in personal and health education and also in science lessons. You will want to get in first, and may be able to remove your child from school classes you do not trust.

Broaching the topics

Each chapter covers a particular aspect of sexual biology: e.g., an overview of the female menstrual cycle, or an explanation of the sexual chemistry of the brain. It takes the form of a conversation between a parent and a child, and may be used in various ways.

- You may want to read the chapter with your child as it stands, stopping to discuss points of interest. The chapters contain a lot of information and many parents break them into smaller units. Sometimes it can be helpful to prepare your child in advance for a topic, describing it in outline on an earlier occasion.

- You may wish to explain a topic in your own words. To help you, each chapter concludes with a list of **Points to Remember** and a **Glossary**, while the **diagrams** have been mounted on their own at the back of the book. These can be cut out and arranged in any order. They can also be printed out from our website (go to the *Sexuality Explained* page of www.alivetotheworld.co.uk). Having prompts helps to keep the conversation going and ensure that you cover all you intend.

- You may want to give a chapter to your child to take away and read. This is particularly appropriate for older children. Even much older children, and adults, can take part.

Setting aside enough time

This is the first imperative. The characters in the book give examples of how to start up a private conversation. Sometimes it just happens, as in the car, when Michael springs a leading question on his father. At other times, Mum and Dad set up the scene, or at least take advantage of a stray reference. Usually the conversation begins while they are doing something, so as not to have to look the children in the eye, and continues at a table where it is easy to pull out paper and pencil or a reference book.

Chapters for girls and boys

The chapters are each composed of a conversation, mostly between a daughter and her mother, and a son and his father. The List of Contents explains where daughter/mother text is intended principally for girls and son/father text for boys. However, sometimes girls like to 'look over the shoulder' at life from a boy's angle, and vice versa.

The material covered in the Girls' and Boys' chapters is similar but not identical, leaving you to add in any details 'from the other angle' which you might find interesting. Who uses what is completely up to each family.

Introduction

Children are generally most at ease when it is somebody of their own sex talking to them. If you feel uncertain, you can always ask someone your child trusts to help you.

The advantages of educating children in the home

Children expect their parents to be their mainstays in everything which touches them closely, and they have a right to look to them for guidance in the sexual field. Parents who take an active interest have more influence over their children than any external agent: something which is publicly recognised but sadly rarely encouraged.

It makes sense that a loving parent or guardian can cater for a child's sensitivity and level of maturity in ways which are impossible during a school sex education class. Nobody knows what damage even one child may suffer from inappropriate group teaching. Sex is a modest subject, as adults know. Children's modesty is just as important, and is part of their armoury of self-restraint.

Building up a relationship of trust in sexual matters is a way to see that parents remain the first port of call for all sorts of advice, just as they were in childhood: if parents are thought unable to deal satisfactorily with adult questions, they risk being supplanted in other areas too.

Some parents shrink from the task, believing that schools are better able to teach well. In practice, school classes often use resources which are worryingly misleading and out-of-date. The errors in Michael's knowledge which Dad corrects in Chapter 7 are common ones. In this case, they were all drawn from a well-respected science book which is widely used in English schools. Teachers themselves are rarely given much training in sex education, while the school nurses brought in to support them are not teachers and hardly know the children.

Moral teaching, or lack of it, is another hazard. Sex education was introduced into schools with the specific short-term objective of bringing down teenage pregnancy and sexual infections. It does this by making contraception well known and easily available. No child will come forward for contraception unless sex outside marriage is also seen to be acceptable, and so sex education tends to belittle the seriousness of the sexual act. School classes are also regularly infiltrated with values which ignore the parameters of life's complete cycle.

It is only you, as the parent or guardian, who will be standing by your child into adulthood and beyond. This gives you a long-term vision which is often absent to others. You are also able to impart sexual knowledge with love in a way that casts out the fear of growing up. In doing so you are recognising your child's emerging manhood or womanhood and the conversations you have together can become an important and enjoyable rite of passage.

Alive to the World

This Guide was conceived as a complement to the *Alive to the World* programme of children's character formation. These excellent books teach children how to live life well as a child and so prepare the foundations for a full and happy adult life.

These foundations are built by perceiving, understanding and practising the universal virtues, without which education in sexuality would be in vain. Integrity, with its indispensable elements of justice, honesty, humility, responsibility, compassion and reciprocity, must be learned by example, by explanation and by practice. *Alive to the World*, through the story of Charlie and Alice and their friends, helps children to comprehend and live these values. Central to this vision is the importance of family and commitment, so giving a context for positive sex education without trespassing upon parents' prerogatives.

The books can be used at home or in school as a full personal, social and health programme, for which there are accompanying teacher guides. The storytelling technique quickly engages children's interest and gives lots of points for discussion. The books are published in the UK by Gracewing.[1] For more details, please see the website www.alivetotheworld.co.uk.

Acknowledgments

This Guide was produced with the help of innumerable people, some of whom I can mention here.

First, I would like to thank Sister Nuala O'Connor, RGS, who, many years ago, trained me as a Natural Family Planning Practitioner, and gave me the knowledge from which this work springs. Dr Thomas Hilgers, founder of the Ovulation Method (Creighton Model)[2], has kindly given me permission to use his original diagrams. These have been reworked and transformed for children by Jessie Gillick, whose witty illustrations bring the Guide to life. I am most grateful to the Knights of Malta whose support made it possible to illustrate the Guide.

Special thanks also go to Christine Vollmer, creator of the *Alive to the World* books, with whom I work closely as her UK Co-ordinator. It was from *Alive to the World* that I took the idea of teaching through stories. Mrs Vollmer has been my mentor throughout and her wise advice is always encouragingly given.

Dr John McLean, retired Senior Lecturer in Anatomy and Embryology at the University of Manchester and consultant in Genitourinary Medicine in the Manchester Royal Infirmary, was the model for Dr Peterson in Chapter 9: 'Dad, Is Sex Dangerous?' The private lesson

---

[1] The UK edition of *Alive to the World* has to date books covering ages 8-13. The Spanish edition, under the title *Aprendiendo a querer*, extends from age 6 right through to 18 (see http://www.alafa.org).

[2] This is now known as NaProTECHNOLOGY. For further information, see http://www.naprotechnology.com.

Introduction

he gave me in sexually transmitted infections extended to checking much of my script. I am also most grateful to Dr Anne Carus, Dr Olive Duddy and Dr David Kingsley, who severally read the text at various stages and gave me their guidance.

The Guide has evolved with the help of many fellow parents. I greatly appreciate the suggestions they have given me, and their trying out the Guide in their homes. Especial thanks also go to Lucy McGough, Susan Cooper and Christina Darby, who proof-read the whole.

Throughout the project my husband David has patiently supported me, giving me ideas and cutting me down to size. It is only with his help, and insights from our four children, that I undertook the work, which is dedicated to them with my love.

Louise Kirk
14 February 2013

Chapter 1

Josie and the Cakes

That Saturday, Josie was helping her mum in the kitchen baking cakes. Dad had taken the others out to the park. Josie enjoyed those moments when she and her mum were alone together. It made her feel more grown-up and she tried to behave like it.

"I'll get the eggs out," she said, and one promptly fell splat on the floor.

Her mother saw Josie's face and chipped in quickly, saying, "Bad luck. Here's some paper towel."

Josie busied herself on the floor before remarking, "Beth's mother is having another baby. I wish we could have one."

Her mother laughed. "You don't think this house is full enough as it is?"

The girl stood up, scrunching the paper towel in her hand. "Mum," she asked, "where do babies come from?"

This time it was Mum's turn to be a bit embarrassed. "Well," she began. "You see these eggs?"

"Yes?" said Josie.

"They come from a hen, don't they? If they'd been fertilised by a cockerel when they were still inside the hen there would be little chicks inside. You see, women also have **eggs** deep inside them."

"Really?" said Josie, her eyes widening. "I've never seen anyone laying an egg."

Mum laughed. "People are similar but not quite the same. You may not know it, but you already have inside you all the eggs you will ever have for your own future babies. About half a million of them."

"Half a million?" exclaimed Josie. "But they'd never fit in."

"If they were as big as these eggs, certainly not. But there is a big difference between hens and human beings. Birds' eggs contain all the **genetic material** necessary to create a new bird, …"

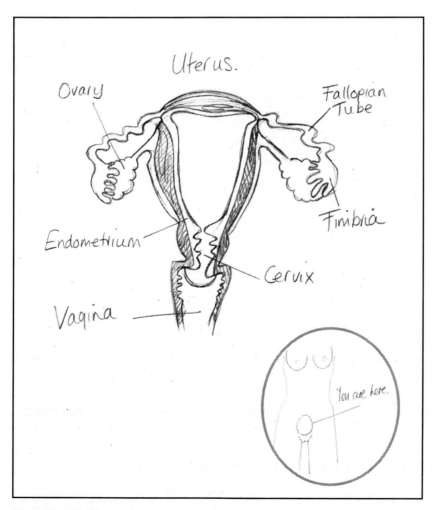

Figure 1: Female reproductive organs

"That's the **DNA**?" interrupted Josie.

"Where did you learn that from? School? Yes, the DNA. But most of a hen's egg isn't DNA. When we eat an egg, what we're eating is the food which a new chick would feed off until the moment it hatches. The hen sits on the egg to keep the chick warm, but it doesn't give it more food.

"With a human mother, the egg contains only the mother's genetic material. That's why it can be so small, minute in fact. Inside the mother's womb—or **uterus**—all the food for the baby is provided by the mother's body.

"Give me some paper from that pad over there, and I'll draw you a diagram of what

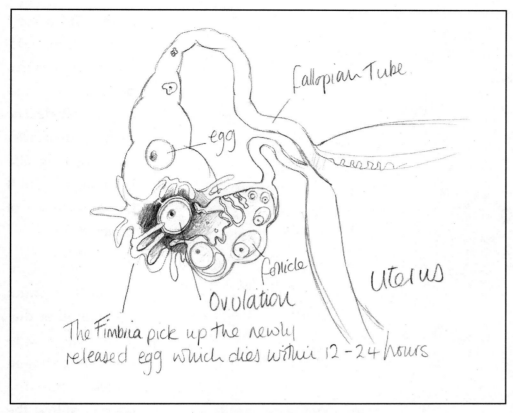

Figure 2: Ovulation and death of unfertilised egg

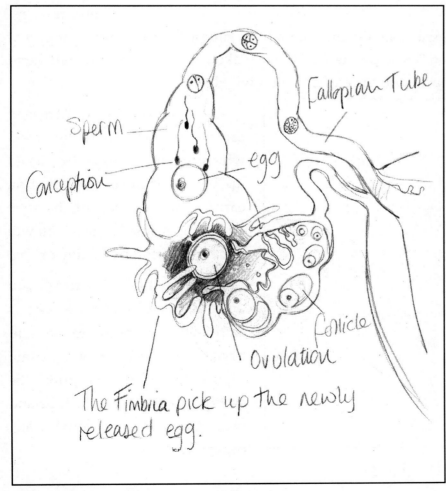

Figure 3: Ovulation and conception

happens each month inside a woman's body. It will begin happening to you in the next couple of years or so, probably when you are 12 or 13, though it can be earlier or later."

Josie brought the paper and sat hunched up on a stool. The cake mixture by this time had long been forgotten. Her mum drew a curious diagram which looked a bit like a sheep's head (figure 1). This she labelled as she spoke.

"All of these organs are right down inside your lower tummy. First of all, you have two **ovaries**, one on the left and one on the right. That's where the tiny eggs are stored. Each month or so, a chemical messenger is sent by the brain to choose an egg, or two if there are to be non-identical twins …"

"Or three?" chimed in Josie.

"Or three, but that's very rare. Anyway, an egg is chosen each month from the

15

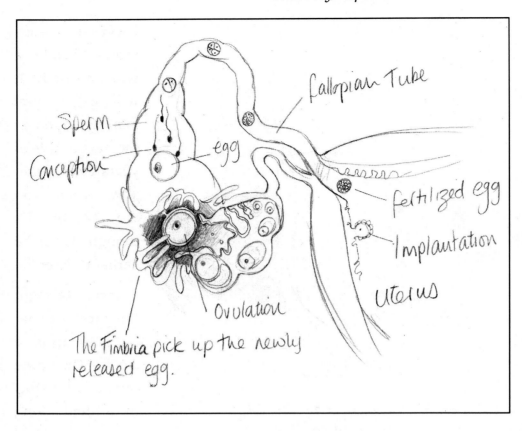

Figure 4: Implantation of the embryo

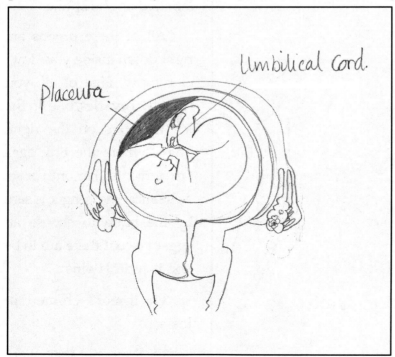

Figure 5: Unborn child at 16 weeks

ovary, it doesn't matter which, and it grows very fast in a protective case called a **follicle**. The follicle grows faster than anything else in the body until it becomes the size of a walnut."

Josie's mum drew another diagram (figure 2) to show the enlarged follicle bursting open to allow the egg to escape. "When it's released from the follicle, the egg also leaves the ovary, to be caught by the feathery fingers at the end of the **fallopian tube**." She pointed to what looked like a multi-fingered hand poised over the ovary and attached to the end of a long wiggly tube.

Her mum continued, "The egg can't move anywhere by itself, which is why it has to be picked up by the **fimbria**. The fimbria carry it into the top of the tube and, unless it's fertilised, it will die there within half a day or so.

"But now I'm going to tell you what happens when the egg is **fertilised** and becomes an **embryo**," she said, drawing again (figure 3). "After conception, the embryo moves along the tube and beds down in the uterus wall" (figure 4).

"How does the egg come to be fertilised, Mum?" asked Josie, looking at the drawing and then up at her mother.

"Well, when a man and a woman love each other very much, they embrace in a special way which causes the man to release **sperm** into the woman's body.

"The sperm look a bit like tadpoles, but they are so small you can't see them, except under a microscope. There are lots and lots of them. Unlike eggs, they have tails and can swim fast through the woman's organs, looking for an egg.

"If they find one, lots of sperm crowd round it. One sperm wins the race and unites with the egg (it seems that the egg chooses that sperm … we don't quite know how … from all the ones surrounding it) to create an embryo and the life of a new baby begins. That's called **fertilisation**, or **conception**, which is a really wonderful moment. It happens here, up at the end of the tube, not far from the ovary. The genes from the sperm and the egg mix together, which is why you're like your dad as well as like me.

Josie studied the diagram. "So conception happens right up here, does it?" she asked, pointing (see Figure 4).

"That's right, up at the end of the tube. Then the **embryo** is squeezed along inside the tube all the way down to the uterus, where the new nest has been made ready. That takes about 5 or 6 days. When the embryo reaches the uterus, it nestles into its inner wall and attaches itself very firmly, which is called **implantation**. The baby grows and grows and nine months later it's ready to be born (figure 5)."

Josie was amazed. "To think all this happened to make me!"

"Yes," said her mum. "That's why you are so unique and so loved. You were made from a piece of me (my egg) and a piece of Daddy (his sperm) and when you think of the enormous choice of cells, both eggs and sperm, from which you happened to be made—an infinite variety—you can see how incredibly unique you are, and your brother and sister. And yet you all come from the two of us and our love for each other."

Josie looked through the diagrams again. Then she said, "But what happens to all those other eggs, I mean the ones that aren't fertilised?"

"Well, very few of the half million ever mature and leave the ovary, and, as I said, most of the ones that do **ovulate** die within 12 or 24 hours after being released."

"It's amazing, isn't it?" said Josie. "Those eggs live inside me all those years, then one of them is chosen and matures, and it goes and dies twelve hours later."

"I hadn't thought of it like that," her mother replied. "But yes, except for the really lucky eggs which are fertilised and become babies, you're right.

"It's not only the eggs which are lost," she added. "If the egg dies, its nest isn't wanted either, so the body gets rid of it. It comes away very gently in the form of blood, a drip or

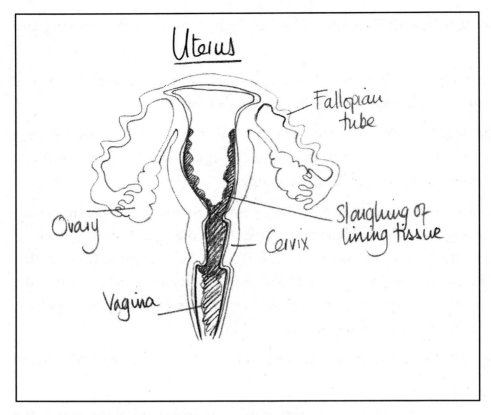

Figure 6: Menstrual bleeding, or period

sometimes a clot at a time. We usually call that a period, and it lasts between 3 to 7 days, though it varies with each woman. "

"What does it feel like?" asked Josie

"The blood itself sometimes tickles a bit, but you can also get a dull tummy ache, or even quite a strong pain in the first day or two. You see the uterus, which is also a strong muscle, is squeezing hard, to loosen the lining. Each woman is different, but it's nothing to be frightened of."

Josie took another look through the diagrams. "Where does the blood actually come out?" she asked.

"It drains down from the uterus through the **cervix**, which is the opening to the womb. The cervix acts like a valve, only letting in and out what it wants to, so that the womb is protected from germs. Then it goes through the **vagina**, which is the tube which connects the uterus to the outside of the body. Your bottom actually has three outlets: one for the wee, or **urine**, one for the pooh, or **faeces**, and the third, which is between them, is the opening of the vagina. That's where the blood comes out—we call it having a **period**. The vagina is also where the sperm come in when the couple embrace in the special way I was talking about."

Josie picked up the empty bowl and began polishing it clean with her finger.

"It must be very messy," she remarked, putting the bowl down. "I mean, all that blood."

Mum went to put the oven on and started gathering things up. "There are special pads you wear inside your knickers," she said. "They're made out of material a bit like disposable nappies, and very absorbent. You change them when you go to the toilet.

"There's nothing to worry about," she added. "The blood drips out quite slowly, though it can be heavier for some women."

At that minute, they were distracted by footsteps outside, and the sound of very familiar voices.

"Quick, Josie! We haven't put those cakes in the oven yet."

"Never mind, Mummy. We have been discussing all sorts of interesting things. And there are doughnuts in the cupboard."

Points to remember

From the moment she is born, a girl has inside her all the eggs she will ever have for her own future babies. There are about half a million of them.

The undeveloped human eggs are so small that they can only be seen with a microscope. Each egg contains all the genetic material that the mother would pass to a child born of that egg.

Babies grow and develop inside the mother who feeds and protects them for nine months until they are born.

Unlike humans and mammals, birds lay their young outside their bodies. The chicks grow and develop inside an egg's protective shell. When we eat an egg, we are eating the food on which the chick would grow.

Human eggs are stored and develop inside an ovary. Each woman has two ovaries.

Each month or so, the brain instructs one of the ovaries to develop an egg inside a protective follicle. The follicle grows to the size of a walnut.

When it is ready, a further instruction from the brain tells the follicle to burst open and release the egg out of the ovary. It is picked up by feathery fingers called fimbria and swept into the fallopian tube which attaches to the woman's uterus.

If there are sperm around, the egg may be fertilised. Conception happens at the end of the fallopian tube. The egg is carried through the tube, implants in the wall of the uterus, and nine months later the baby is ready to be born.

Most months, the egg remains unfertilised. It dies in the tube within 12-24 hours.

The uterus then sheds the lining it has prepared as a 'nest' for a possible baby.

This comes away in the form of blood and drips out of the vagina. The bleeding usually lasts for between 3-7 days and is called a period (or menstruation).

Girls usually start their periods at the age of 12-13, though it can be several years earlier or later than this.

GLOSSARY

Cervix	Neck of the uterus which opens and shuts to control access to the uterus and keep germs out.
Conception	The start of new human life, when the egg and the sperm fuse to form a new cell with its own identity (DNA). Conception occurs towards the ovary end of the fallopian tube.
DNA	Deoxyribonucleic acid. DNA is present in the nucleus, or control centre, of each of the cells from which living beings are made. Inside the DNA, there are tiny **genes** which contain instructions for how cells are to grow and behave. Each person's DNA is unique, and every cell of our body is marked with that unique DNA, except for our eggs or sperm which, remarkably, each have their own DNA.
Egg	Female reproductive cell, which, if fertilized by the male sperm, can develop into new human life. Girls are born with about half a million tiny eggs already stored in their ovaries.
Embryo	Baby in the first eight or so weeks of life after conception.
Endometrium	The scientific name for the lining of the uterus, which thickens each cycle and comes away in a period.
Faeces	Waste matter from food which is expelled from the body through the anus.
Fallopian tube	Small tube which takes the egg towards sperm and sperm towards the egg. Conception, if it occurs, happens in the outer part of the tube. Without sperm, the egg dies in the tube.
Female reproductive organs	Organs inside a woman's lower tummy which, when they act in harness with a man's reproductive organs, have the potential to give life to a baby.
Fertilisation	Process by which the sperm becomes one with the egg to form a new human being. Fertilisation can take up to a day to complete.
Fimbria	Feathery 'fingers' at the end of the fallopian tube which catch the egg from the ovary and carry it into the tube. The egg has no means to move on its own.
Follicle	Protective case which grows up round the eggs as they mature. The main follicle grows faster than anything else in the body, and becomes the size of a walnut before breaking open the ovary to release the egg.
Genetic material	Collective term for the package of genes which govern the way cells behave. Genetic material is stored in DNA.

Implantation	The process whereby the embryo roots into the wall of the uterus, where it will feed and grow. The lining of the uterus thickens each month to prepare a 'nest' for a possible baby.
Ovary	Organ the size of a Greek olive which houses and matures the eggs. Each woman has two ovaries.
Ovulation	Moment when the egg is released by the follicle from the ovary.
Period	When the lining of the uterus is flushed out because there is no pregnancy. Women experience this as a periodic bleed from the vagina. It usually lasts for several days and takes place every month or so, which is why it is also called **'menstruation'**.
Placenta	Organ attached to the wall of the uterus which transfers oxygen and nutrients to the growing baby, and takes away waste products. Blood flows through the placenta between mother and baby while keeping the two circuits independent. Babies can thus belong to different blood groups from their mothers.
Sperm	The male reproductive cell. It has a head, which contains the man's DNA, and a tail, which enables it to swim through the uterus and up the tube. Only one sperm, out of many sperm that the man releases, can unite with the egg.
Umbilical cord	Attaches the baby to the mother. As the baby grows bigger, it needs to be able to exercise. The umbilical cord gives it freedom of movement away from the placenta and the uterine wall. By 16 weeks, the baby is largely formed and just has to grow bigger and stronger.
Urine	Liquid which removes waste products from the blood.
Uterus	An organ made of strong muscle where the baby develops and is nourished before birth. It is also called the **'womb'**.
Vagina	Tube which connects the uterus to the outside of the body.

Chapter 2

Josie Lends a Hand

They were sitting together in the sunshine, taking out curtain hooks. It was just the weather for washing and Josie had volunteered to help.

"Mum," she began. "You remember you once told me about having **periods** and all that stuff? Well, I was trying to explain it to Beth, but I couldn't remember it all. Can you tell me again, please?"

"Josie darling," her mum replied. "I know you want to help your friend, but it really is for Beth's mum to explain those personal things to her."

"But she hasn't, and Beth wants to know. She's really worried about getting a period and bleeding all over the place. I told her it wasn't like that, so she asked me what I knew."

"Well, I suggest you tell Beth to go and ask her mum. She'll be able to explain everything to her. Her mum may be a bit shy of bringing the subject up and be really glad if Beth asks her. Anyway, what have you told her?"

"I started by telling her why we have periods," Josie replied. "I told her that it's a sign that a girl is becoming a woman. Each month a new nest is made inside the **uterus** for a baby to be nurtured. Of course a woman doesn't have a baby by herself. She can only have one if she's had

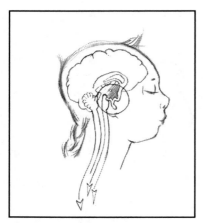

intercourse with a man. Most of the time there's no baby so the nest isn't wanted. The body gets rid of it in the form of drops and clots of blood. The bleeding usually lasts about 3 to 7 days, heavier at first and then very light, and that's what the period is."

Josie's mum smiled. "That's a very good explanation. But did you explain about the **egg**?" she enquired, putting down one curtain and picking up the next.

"I told her that all the eggs we'll ever have are in our **ovaries** before we're born. She was really impressed at that. And she didn't believe me when I told her that, although a lot die off, a girl still has about half a million at puberty. 'Half a million periods?' she said. But I told her that only 400 or 500 of them ever mature," Josie finished.

"Did you tell her about **ovulation**?" her mum asked.

"That rings a bell, Mum, but I can't remember the details."

Josie's mum put down the curtain and went inside to get some paper. "Can you wipe the table down, dear?" she called over her shoulder.

Drawing a chair up, she settled herself down and

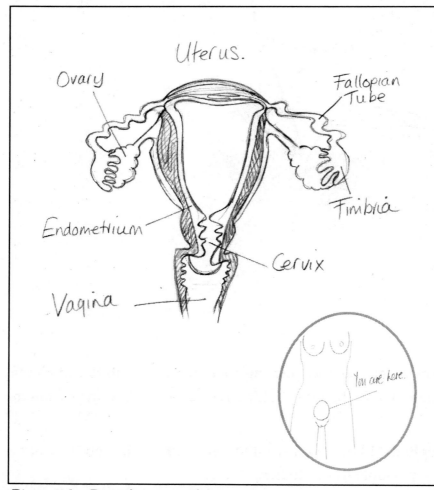

Figure 1: Female reproductive organs

began drawing a familiar shape. It included the **uterus**, the **endometrium**, the **ovaries**, the **fallopian tubes**, the **cervix** and the **vagina**.

"I should really add in the woman's brain as well," she said. "A lot of the instructions which control the reproductive cycle come directly from the brain."

She quickly sketched in a head with arrows down towards the other organs (figure 1). "I like to think of the brain as our main sexual organ. In addition to sending crucial instructions, it's at the centre of who we are as women and men.

"There," she said with satisfaction. "Now you can see the **uterus**, the two **ovaries**, one on each side, and the **cervix**. The cervix is really important. It acts like a gateway to the uterus and controls everything that comes in and goes out.

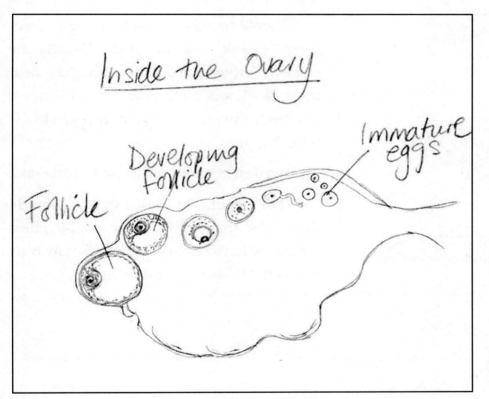

Figure 2: The development of the egg in the ovary ready for ovulation

"Let's start with the ovaries. I'm going to blow one up so we can pretend we're looking inside it (see figure 2).

"All those little dots are the undeveloped eggs. Of course, in real life they're too small to see without a microscope. Every month, the brain instructs one of the ovaries to bring an egg to maturity. Can you remember how the ovary is chosen?"

Josie looked puzzled. "I imagine they alternate?"

"You might think so, but actually it's more haphazard than that. The egg can come from either of the two.

"By the beginning of the **menstrual cycle**, several eggs inside one of the ovaries are beginning to mature inside the shells called **follicles**. The follicles don't just protect the eggs—they also produce a **hormone**, or chemical messenger, called **oestrogen**. This tells the uterus to rebuild its lining, or **endometrium**, and make a new nest again.

"After about eight days, one of the follicles grows so rapidly, it takes over from the others. Soon it fills the whole ovary. You can imagine the amount of oestrogen it produces. When the level in the blood is high enough, the oestrogen acts on the brain, which in turn tells the main follicle, or the **dominant follicle,** to release its egg. Out the egg bursts, with such energy that it tears the ovary open and pops out of that too, into the **abdomen**. And that's called **ovulation**."

"And I know what comes next," Josie went on enthusiastically. "When the egg leaves the ovary, it's picked up by those **fimbria** at the end of the tube and then it's carried down the tube toward the uterus. And that's where the baby develops."

"There'll only be a baby if the egg's fertilised!" Josie's mum laughed. "Usually the egg isn't fertilised and it dies about 12—or at most 24—hours after ovulation so it never reaches the uterus at all. And then you know what happens."

"The nest in the womb is shed," Josie said.

"That's right. The period begins some 14 days after ovulation. And can you guess where the hormones come from that instruct the uterus to shed its lining?"

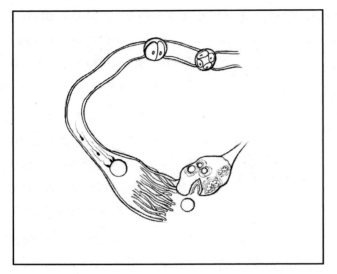

Figure 3: Ovulation and conception

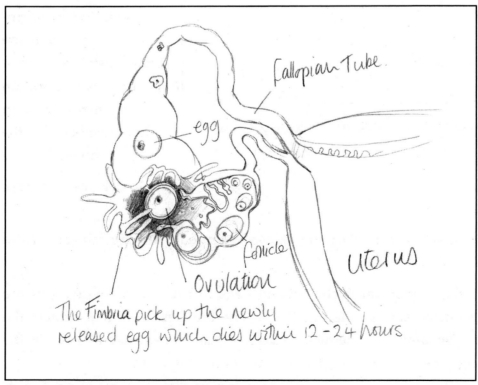

Figure 4: Ovulation and death of the egg

"The brain?" Josie volunteered.

"No," Mum replied. "Try again."

Josie looked hard at the diagrams. "Well, there's nothing much left in the ovary except the empty follicle" (figure 5).

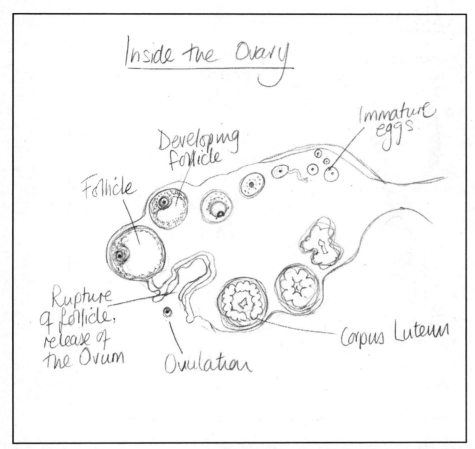

"But it isn't empty," said Mum. "After ovulation, the burst follicle fills with blood and becomes what's known as the **corpus luteum**—the Latin for 'yellow body'—because it looks yellow. This corpus luteum releases another hormone called **progesterone**, which is essential for keeping the thickened endometrium."

"Doesn't the oestrogen do that?" Josie asked.

Figure 5: Full cycle of events in the ovary; the empty follicle becomes the corpus luteum

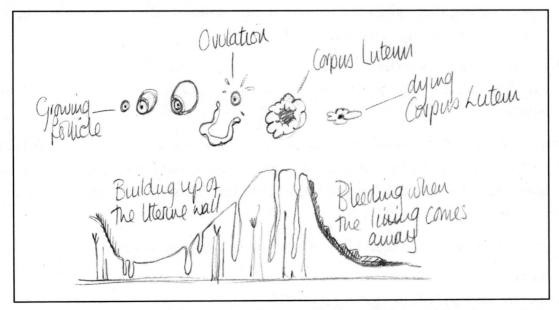

Figure 6: Cycle of events in the uterus with ovulation cycle shown above

"The corpus luteum goes on producing some oestrogen, and together they maintain the nest, but the progesterone also stimulates the endometrium into producing the rich juices which would feed a baby."

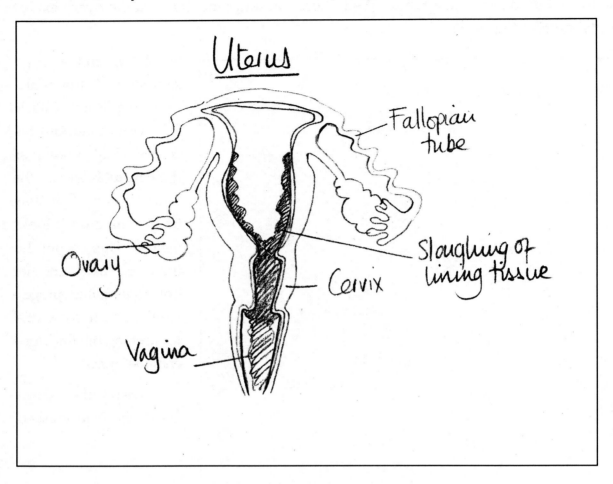

Figure 7: Menstrual bleeding, or period

"So you're saying that, when the corpus luteum disintegrates, there's nothing there to produce any more hormones, so the nest falls away?"

"That's right. If there's no baby, there's nothing left to sustain the endometrium and so it falls away. It's nature's way of cleaning up and starting again with a fresh nest for each possible child."

Mum returned to the curtains while Josie sat still looking at the diagrams. "Mum," she asked. "You know you told me that a man and a woman have a special embrace, and the man releases sperm into the woman's body? Beth told me that the man's penis enters the woman's vagina. Is that true?"

"It's sounds rather odd, doesn't it?" Mum replied. "But yes, it's true."

"And that's what intercourse is? And making love?" Josie went on. She didn't wait for an answer before saying, "Mum, doesn't it hurt?"

The Menstrual Cycle

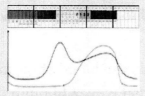

Periods are called 'menses' or 'menstruation' from the Latin word for 'month'. They are the most noticeable symptom of the female reproductive cycle, which also includes the release of an egg, the building up of the lining of the uterus, and the release of mucus from the cervix.

The amount of loss varies, but is typically no more than a couple of tablespoons over 3-7 days

The time of first period (**menarche**) also varies. Many girls start at age 12 or 13, and most between the ages of 10 and 15, but it can come as early as 8 or as late as 17. **Menarche** can be expected a couple of years after the breasts begin to bud.

Ovulation may begin before the first period, or some time later. Once periods become regular, it is a sign that ovulation has also got underway.

It takes a year or two for the body to settle into a regular menstrual cycle. A girl may have a period and then nothing for several months. As she gets older, the cycle becomes more predictable, with an average 21 to 34 days between the start of one period and the start of the next. It helps to jot down dates in a diary.

Girls vary enormously in the symptoms they get, and some feel very little. A hot drink and warm bottle are always comforting, and Ibuprofen or Paracetamol will normally relieve tummy cramps.

Mum shook her head. "No, because the body is very clever and is made for this. When a man and woman are married and have given themselves totally to each other, intercourse is like a very special cuddle and each of their bodies gets ready for the other. This's why it's important to be totally committed to each other. Your vagina and cervix will go on growing and developing until your late teens, so, before that, making love can be quite uncomfortable for a girl. Premature sex is not the same as the real thing."

She began gathering up the curtains. "Anyway, I must get these curtains into the machine before lunch. I promise we'll finish off our conversation another day. Apart from anything else, I haven't yet shown you where I keep **sanitary pads**—you ought to know in case you begin a period when I'm not around."

"Oh, yeah, Mum," Josie broke in, also getting up. "Beth asked me about **tampons**. What are they, and when do you use them and when do you use pads?"

"I'm glad you asked that, because that's one thing you can pass on to Beth. Pads stick inside your knickers and catch all the blood as it comes out. Tampons are made of very tightly compacted absorbent material and they sit inside the vagina. They catch all the blood inside you. They're very convenient, especially if you're swimming or playing sport, but they can also be dangerous. I'll explain why quickly.

Hygiene

The hormones which control the menstrual cycle also act on the sweat glands.

Once puberty starts, young people need to wash their hair more often and start using deodorant.

"During a period, your body is flushing out material it doesn't want. If you use a pad outside the body, all that unwanted material can drain away naturally in a healthy way. Also, it's easy to remember to change the pad because you see the blood.

"With a tampon, the blood stays inside you and can't escape in the normal way. Although the tampon absorbs it, the blood isn't being removed from the body. Bacteria thrive in the rich menstrual blood and—just occasionally—infection sets in. This can cause **Toxic Shock Syndrome**. It's rare, but if it happens it's very serious and can even kill you.

"Anyway, I'm not going to suggest that you use a tampon until your periods are very well established and your body is formed. The thing to remember is: only use tampons when it's really necessary, and change them frequently—at least every 4–6 hours, depending on the flow. Never wear them at night, and avoid them during the lighter days of bleeding—the temptation not to change them often enough is too great."

Mum looked down at her watch. "Hey, look at the time!" she exclaimed. "You're now going to have to help me get the dinner!"

Points to remember

The brain is our main sexual organ. It has overall control of the menstrual cycle and is also at the centre of who we are as men and women.

The follicle which protects the maturing egg gives off oestrogen, and the empty follicle, known as the 'corpus luteum', gives off progesterone.

The oestrogen is responsible for building up the lining of the uterus, called the endometrium, in which a fertilised egg will implant.

The progesterone maintains the endometrium and stimulates the production of the juices which would feed a baby.

When the corpus luteum finally collapses, the production of oestrogen and progesterone rapidly drops. Without hormonal support, the endometrium falls away, causing a period.

Nature has adapted the bodies of a husband and wife to combine at intercourse in a close cuddle in which the sperm enter the woman's abdomen through the vagina.

Premature sex is not the same thing and can be uncomfortable for a girl whose vagina is not yet fully formed.

Sanitary pads inserted in the knickers collect the menstrual blood outside the body. They are made of material resembling disposable nappies and come in a range of absorbencies.

Tampons absorb the blood inside the vagina. They are convenient, especially during sport or swimming, but prevent the body from flushing out material it doesn't want. Bacteria thrive in the rich menstrual blood and, very occasionally, can cause Toxic Shock Syndrome.

Tampons should only be used when really necessary, and changed frequently. It is better to avoid them in the younger teenage years when the vagina is not yet fully mature.

GLOSSARY

Abdomen	The cavity in the lower part of the body, below the diaphragm, which houses the major organs of reproduction and digestion. The whole cavity is lined by a membrane called the **peritoneum**.
Cervix	Neck of the uterus which controls the access of sperm to the female organs. Mucus from the cervix either lets the sperm through or blocks their entrance.
Conception	The start of new human life when the egg and the sperm fuse to form a new cell with its own identity (DNA). Conception occurs at the ovary end of the fallopian tube.
Corpus luteum	Name given to the collapsed follicle, meaning 'yellow body' in Latin. The corpus luteum produces the progesterone.
Dominant follicle	Follicle which grows fastest, and takes over from the other follicles. It bursts at ovulation to release its egg.
Egg	Female sex cell which contains all the mother's potential for giving life to a child. Each cycle, a dominant egg develops within a follicle to be released into the abdomen at ovulation. The egg is picked up by the fimbria at the end of the fallopian tube. Unfertilised eggs die within 12 or 24 hours after ovulation. Also called by the Latin name **ovum** (**ova** in the plural). Occasionally two eggs are released at ovulation, but always within hours of each other. If fertilised, these can become non-identical twins.
Embryo	Term given to the newly conceived baby in the first eight weeks or so of life.
Endometrium	Inner lining of the uterus which thickens each month to prepare an hospitable environment in which an embryo can implant and grow.
Fallopian tube	Tube (also called an **oviduct**) of about 8 to 10 cm in length, which takes the egg from the ovary, and any sperm towards the egg. Named after the great sixteenth-century Italian anatomist, Gabriele Fallopio.
Fertilisation	The process by which the egg and sperm fuse to become a new human being.
Fimbria	Finger-like extensions of the fallopian tubes which catch eggs from the ovaries.
Follicle	Shell-like structure in which the egg matures inside the ovary. The developing follicle gives off the hormone oestrogen.

Hormones	Chemicals which circulate through the blood-stream, giving instructions from one organ to the next. Hormones can originate in the brain, or in other organs such as the ovary. They often trigger each other in relay.
Intercourse	The act of making love, when sperm pass from the man to the woman.
Menarche	A girl's first period, or the start of menstruation (and ovulation).
Menstrual cycle	Cycle of reproductive events in the woman's body which, by tradition, is said to begin with the first day of the period and end with the last day before the next.
Oestrogen	Female sex hormone which tells the brain when the egg is mature, instructs the lining of the uterus to thicken, and controls the production of fertile cervical mucus.
Ovary	Organ shaped like a Greek olive, which houses and matures the eggs. Women have two ovaries which contain, even before they are born, all the eggs they will ever have (about 500,000). Each cycle, an egg is chosen from one of the ovaries (not necessarily alternately) to mature and ovulate.
Ovulation	Process by which the mature egg is released from the follicle and out of the ovary.
Penis	Male sex organ which releases sperm into the vagina.
Period	Bleeding from the vagina as the endometrium falls away in the form of drops and small clots of blood. Periods normally last about 3-7 days, starting more heavily and tailing off. They are a regular feature of a woman's life, occurring within a couple of weeks of each egg ovulating.
	The pattern that periods take can vary enormously from one woman to the next, but they most commonly happen monthly, which is why they are also called **menstruation** or **menses**. A girl generally has her first period about two years after her breasts begin to bud.
Progesterone	Female sex hormone which stimulates the glands in the endometrium to produce nutritious fluids, and the glands in the cervix to produce barrier mucus.
Sanitary pad	Press-on towel worn as lining to knickers to absorb the blood released during a period. Pads come in different shapes and thicknesses. Most modern pads are made of material similar to disposable nappies.
Sperm	Male sex cell, through which a man fathers a child.

Tampon	Tightly compacted stick of absorbent material which sits inside the vagina, catching the blood from inside. Tampons are convenient, especially during sport, but can be dangerous if left inside the body for more than a few hours (4-6 hours is usually the longest time recommended).
	Menstrual blood is rich in nutrients and easily harbours bacteria: it is designed to flow out of the body. Blocked in by a tampon, bacteria can give rise to Toxic Shock Syndrome.
	Girls whose bodies are not yet fully formed are usually advised against using tampons and should consult before doing so. It is important to use tampons of the correct absorbency for the flow.
Toxic Shock Syndrome	Rare but potentially fatal disease caused by toxic bacteria.
Uterus	A muscular, pear-shaped organ, in which a baby develops and is nourished before birth.
Vagina	Muscular canal which leads from the uterus to the outside of the woman's body.

Chapter 3

Josie's Next Lesson

"Mum," Josie said one day, pushing back her chair. "How will I know when I'm about to have a period?"

Josie was sitting at the kitchen table, supposedly doing her homework. Her mum looked up from the sink at the unexpected question. "You won't know first time round," she replied, "but don't worry. At first you're unlikely to get much blood, and it may only be a small amount of brown staining in your knickers." She stopped to pick up a plate, then continued, "I remember when that happened to me and I thought 'What's this? This isn't blood!' In fact it's dried blood. Once your periods get underway, the discharge becomes fresh and red. Even then, it's a discharge rather than a sudden whoosh—it's not like when you go to the toilet."

"It must be a bit embarrassing, though, if you suddenly get blood appearing when you don't expect it!" Josie exclaimed, shutting her exercise book and putting down her pen.

"Well, it certainly helps to be tuned in to what's going on," her mum replied. "Periods can be quite erratic when they first start, but once they get going you'll get to recognise your own pattern. That helps to get you prepared, so you aren't caught unawares. If you mark off on a calendar when each period begins, you'll know when to expect the next." She paused for a moment, then added, "Some people are more regular than others, but even if you don't have a standard pattern the body has some give-away signs, if you know how to look out for them. I don't think I told you about the **mucus** last time, did I?"

"Mucus, like when you have a cold?" Josie asked.

"I obviously didn't. Yes, it's a bit like when you have a cold, but it comes as a discharge from the vagina, like the period. And it happens around the time of **ovulation**. If you see the mucus, it's a telltale sign that you can expect your period in about a fortnight. Once you get to know yourself, you'll know the timing more accurately."

"What's the mucus for, Mum? I mean, I don't suppose you just get it."

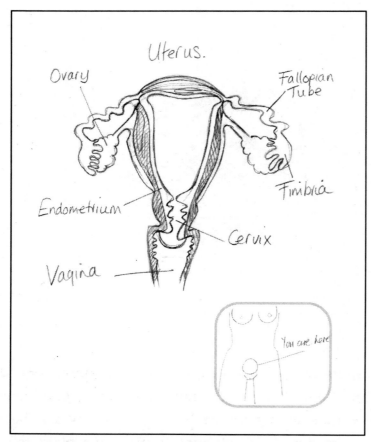

Figure 1: Female reproductive organs

Her mum laughed. "Look, Josie. You're meant to be doing your homework."

"It's OK, Mum, I've finished it all. Promise."

Her mum wiped her hands and came over to the table. She looked through what Josie had being doing before shutting the books and replying, "You were asking why we have mucus. If you want me to explain properly, I'd better show you with some diagrams. Have you still got the ones I drew last time? I'll get one or two from a book I have, too."

Josie ran upstairs and brought them down.

"Here you are, Mum," she said, sitting down expectantly beside her.

"Right," Mum replied, smoothing them out and choosing the diagram which showed all the reproductive organs together (figure 1). "Now, you'll remember what the **uterus** looks like, with the **ovaries**, the **tubes** and the **vagina** down here."

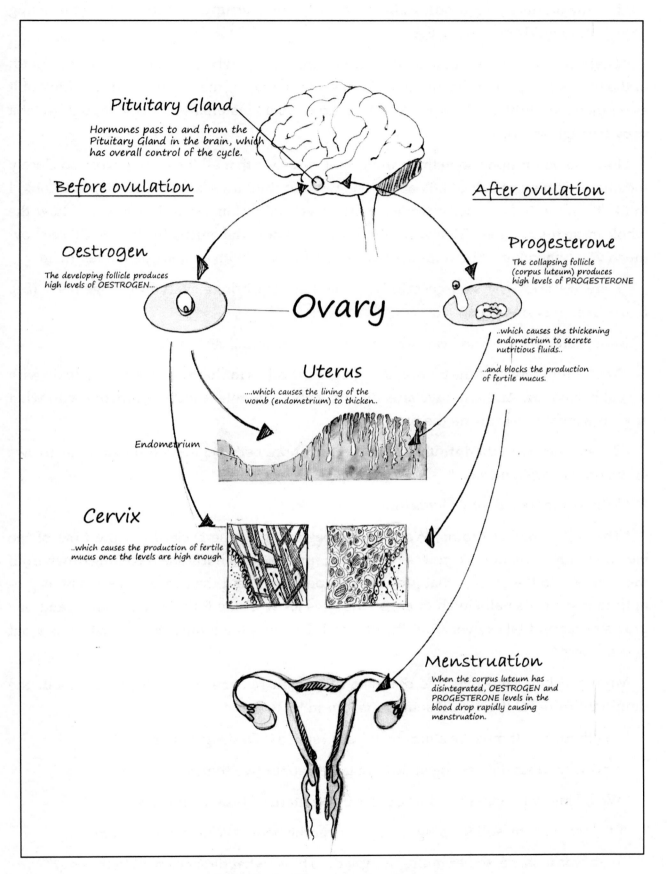

Figure 2: The menstrual cycle

"Is it the uterus which produces the mucus?" Josie interrupted. "I remember you saying it gives out liquids to feed the egg."

"That's a clever guess, but actually it comes from the **cervix**, which is here, at the mouth of the uterus. It's part of the uterus and links it to the top of the vagina. You can think of it as an inch-long tube made of lots of really expandable material—stretchy enough to let a baby through at birth.

"But you've hit upon something important, which is that all the organs work so closely together that it's difficult to talk about one without thinking about the others, too. Look, I think it's going to be easier if we leave the cervix for a moment and I show you how the whole cycle fits together. That way you'll really see how the mucus fits in, and I'll see how much you remember." She went upstairs and came back with a book open in her hands.

"Here we are," she said cheerfully. "This diagram brings everything together." (See figure 2 on previous page.)

Josie peered over. "And I can see it starts with the brain," she said.

"Yes," Mum replied, "the book and I agree. The brain's in charge of everything to do with sexual behaviour. And what we already know about its role is little, compared with what scientists are discovering all the time.

"So, we start with the **pituitary gland** in the brain. Let's begin by following the arrows down on the left-hand side."

"Where it marks '*Before Ovulation*'?" Josie asked.

"That's it," replied Mum. "We're at the beginning of the cycle, from the time of the bleeding until ovulation. So you see an arrow representing a hormone is coming down from the pituitary to the ovary—that grey oval shape there. And the ovary responds by beginning to mature a small clutch of eggs. The eggs grow inside follicles, don't they, and one grows faster and takes over from the others. It becomes the **dominant** one, which is what you see here."

Mum put her hand over the diagram. "Can you remember what's special about the **dominant follicle**? It gives off something, doesn't it?"

Josie thought. "It must be a hormone," she replied. "**Oestrogen**, that's it."

"And what does the oestrogen do? We talked about two things last time."

"Well, I know it makes the lining grow in the uterus," Josie volunteered.

"OK," said Mum, still keeping her hand over the book. "What else does it do?"

There was a pause, so her mum prompted, "Think, what happens next to the egg?"

"Ovulation," said Josie. "Oh, I get it. The oestrogen tells the brain that everything's ready, and the brain sends down a hormone which makes the follicle burst."

"Well done. So we have a two-way arrow going to and from the brain." Mum moved her hand to point to the arrow. "Now look at what's going on down here in the **cervix**. When the oestrogen levels are high enough, and they have to be high, the cervix begins to produce **mucus** which lets the sperm in. It's called the **fertile mucus** and here it is in the diagram. Can you see all the little channels with the sperm swimming up through them?

"That covers the left-hand side of the diagram. Now, let's go to the right-hand side."

"Where it marks '*After ovulation*'," Josie added.

Again Mum pointed to the double arrow to and from the brain. "The hormone from the brain which tells the egg to ovulate also acts on the empty follicle," she explained. "You have to remember that the follicle is quite big, isn't it, about the size of a walnut, so there's a lot of collapsed follicle left when the tiny egg, which is about the size of a full-stop, leaves.

"Can you remember what the empty follicle's now called?" Mum asked.

"I remember it meant 'yellow body' in Latin," Josie replied, pushing Mum's hand away, and reading out: "It's the **corpus luteum**."

"Well done," Mum said. "The corpus luteum continues to produce some oestrogen, but under the instructions of the brain it also begins to produce high levels of **progesterone**. Can you tell me what that does?"

Josie replied, "Well, in the uterus the juices are produced which could feed a fertilised egg, and in the cervix I suppose the mucus changes to infertile?"

"Good. I'll tell you more about it in a second. But now think about the period. Can you tell me what causes that to start?"

"No," Josie replied, honestly.

"I'll give you a clue. The corpus luteum has gone on producing oestrogen and progesterone, but what's happening all the while to the corpus luteum?"

"It's disintegrating," Josie replied. "Oh, I remember!" she exclaimed suddenly. "The corpus luteum disappears, and so the oestrogen and progesterone both stop too! And then there's nothing to support the lining any more, and so it all falls away. It's so clever."

"The other thing to point out is that the **first part of the cycle**, the time when the follicle is building up before ovulation, can vary a lot in length. For most women, it's about **two weeks**, but it can be anything from **one** to **six**. But the **second part of the cycle**, after ovulation, when the corpus luteum disintegrates, is regular. It's usually about **13 days**, though it can be a few days more or less. The point is that even if your whole cycle varies from month to month, you can still get a good clue about when the next period is due from watching the mucus."

"So, the time it takes for the eggs to be chosen and mature varies, but the time it takes for the corpus luteum to break up is pretty regular?"

"That's it, and each woman soon learns her own pattern."

"You were going to tell me about the mucus," Josie prompted.

"Yes, I'm just coming on to that," her mum replied, taking some more paper. "I'm going to draw the cervix by itself, a bit bigger (figure 3). I told you it had to be able to stretch big enough to let a baby through. That's why it's got all those wiggly folds. They're called **crypts,** and inside the crypts there are hundreds of glands which produce the mucus. They don't produce just one type—there're lots of different kinds. The scientists are still discovering quite how many there are and what they all do. Some you see on the outside of your body, and some you don't. Anyway, I'm going to keep it very simple and talk about the two basic types: one is fertile and lets sperm through into the uterus, and the other is infertile and blocks it out.

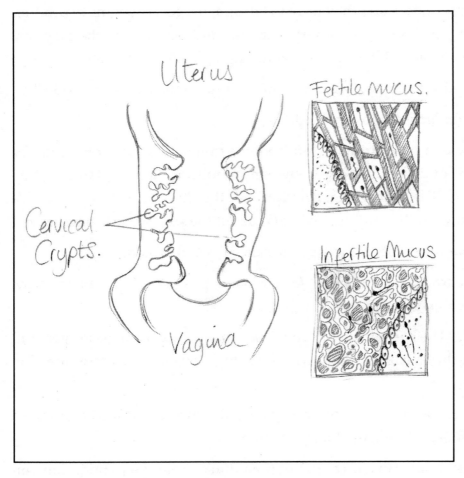

Figure 3: The cervix and its mucus

"Most of the time the cervix produces mucus which looks like this." She pointed to her right-hand diagram. "Can you see that it is made up of blocks designed to keep things out? That gives some protection against germs, but its main purpose is to prevent any sperm getting through if the woman has had intercourse."

"Oh," said Josie. "Does that mean that most of the time you can't have a baby?"

"You can't have a baby most of the time anyway, can you? You can only have a baby when there's an egg ready to be fertilised, which only happens at ovulation. From what I've already told you, how long do you think the fertile period lasts?"

Josie thought hard.

"I'll give you a clue," her mum ventured. "How long can an unfertilised egg live for in the fallopian tube?"

Figure 4: Fertile cervical mucus under a microscope

"Up to 24 hours?" Josie asked.

"Well done. Up to 24 hours but usually nearer 12. So that means that conception, when the sperm and the egg join together, can only take place within 12–24 hours each menstrual cycle. But nature has extended the period of time when an act of intercourse can lead to conception by something else. What do you think that could be?"

Josie looked really puzzled at this. "Can't think!" she said after a bit.

"If you want to be sure to catch a bus, and you know roughly when it's due, what do you do?" her mother asked.

Josie shrugged and said, "Arrive early and hang around, I suppose."

"If sperm arrive early, that's exactly what they do. They hang around and wait for the egg. But they're only able to do it when the mucus has changed to fertile." She produced a photograph of cervical mucus from the book (figure 4). "Can you see all those channels? This is the fertile mucus, and the sperm are able to swim through it, even being fed as they go. Now you can see how necessary the mucus is. The sperm can stay alive in it, waiting for the egg, for about 3–5 days."

Josie looked again at the photo. "Wow, Mum. It's amazing! Isn't nature clever?"

"You should see how much the hormones swing to bring it all about," Mum added, sketching again. "The levels in the blood from one period to the next would look something like this"(figure 5).

"When the oestrogen level's high, the mucus flows and the cervix becomes accessible; when it's low, the cervix closes up. There's even a telltale sign on the outside of the body—you can see the fertile mucus clearly, though it's important to know exactly what you're looking for, because you can also get other vaginal discharges."

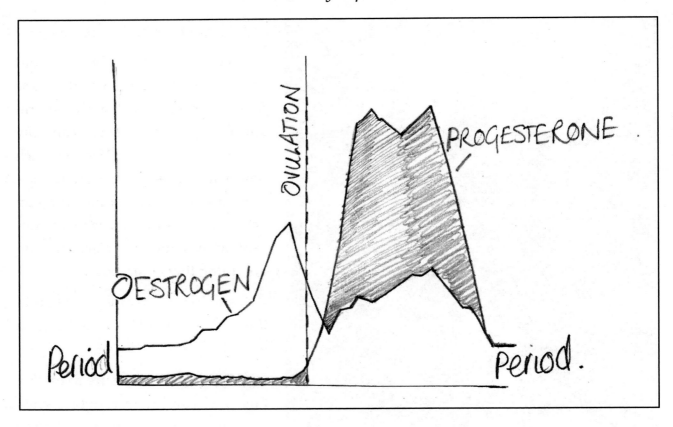

Figure 5: Female hormone cycle

"Mum, if you can only start a baby for such a short time each month, why does everybody talk about using contraception?" Josie asked.

Her mum met Josie's inquiring look. "There are various reasons, but I suspect the main one is that most people don't know how the body works and how easy it is to read its language. Your dad and I only learnt about it recently and we were so impressed that we made it our business to find out more. Now, we want you—and your brother and sister—to know and respect the full beauty of your bodies from the beginning."

Josie's mum paused and smiled. "Your generation is much luckier than mine. You see, a lot of the science has been discovered comparatively recently and it hasn't been widely taught."

She looked down again at the diagrams in front of her. "When you're older I'll teach you how to recognise all the signs of your menstrual cycle. But for today it's enough to remember that there are two main types of mucus. One nourishes and helps the sperm along, and the other blocks it. There's another big difference. The barrier mucus stays where it is, inside you. You won't be aware that it's there. But the stringy mucus—the mucus which looks after the sperm—drips down through the vagina and is clearly visible on the outside of the body. It appears as a sticky discharge, a bit like white of egg. Sometimes it's like a gluey white lump. You'll come to recognise it. When I was young, nobody told me about it. I remember seeing it and thinking there must be something wrong with me! I thought I must have tape worms (I didn't know what they were either)! When you see it, you'll know it's a pretty good clue that you can expect a period in about a fortnight's time."

Josie looked up at her mum and gave her a big hug. "Thanks, Mum," she said. "You know, you're the best mum in all the world!"

Points to remember

Periods can be erratic when they first start. Once they become established, women get to know their own pattern.

Mucus discharge round the time of ovulation indicates that a period is likely to happen in about 13 days' time. It usually lasts for several days, but can be more restricted in less fertile women.

Even when periods are irregular, the phase between ovulation and period usually remains stable.

The variants and functions of cervical mucus are still being studied, but the mucus can be broadly defined as fertile or barrier.

The fertile mucus appears round the time of ovulation and is obvious to the woman once she has been taught what to look for. Women can also experience other vaginal discharges.

All of the woman's reproductive organs work in close synchronisation under the control of the brain.

Hormones from the pituitary gland in the brain start the cycle by stimulating the dominant follicle into producing oestrogen. The oestrogen in turn builds up the endometrium and, when sufficiently plentiful, starts the production of fertile mucus.

Another hormone from the pituitary prompts ovulation, and gets the corpus luteum to produce progesterone. The progesterone in turn prompts the production of nutritious juices in the endometrium, and gets the cervical crypts to produce barrier mucus.

When the corpus luteum disintegrates, hormonal support falls and the endometrium pulls away from the uterus, causing a period.

The fertile mucus can prolong the life of the sperm for up to 3–5 days, allowing them to linger in the female tract awaiting ovulation.

GLOSSARY

Cervix	Neck of the uterus which acts as a valve. Sperm are only allowed in when the mucus it produces is of the fertile type.
Corpus luteum	Name given to the collapsed follicle. The corpus luteum produces both progesterone and oestrogen.
Crypts	Pockets found in the lining of the cervix which produce both the barrier and the fertile mucus.
Dominant follicle	Largest follicle to develop in any one cycle, taking over from others in the ovary and causing ovulation when it releases its egg. The dominant follicle gives off oestrogen.
Mucus	Gel-like substance with varying water content, produced in the cervical crypts. Barrier mucus prevents sperm from entering the reproductive tract. Fertile mucus is produced round the time of ovulation and helps sperm on their way towards the egg.
Oestrogen	Female sex hormone which tells the brain when the egg is ready for ovulation. It instructs the lining of the uterus to thicken, and controls the production of fertile cervical mucus. Also spelt **estrogen**.
Ovulation	Process by which the mature egg is released from the follicle and out of the ovary.
Pituitary gland	Small gland found at the base of the brain which secretes the hormones which control the menstrual cycle.
Progesterone	Female sex hormone which stimulates the glands in the endometrium to produce nutritious fluids; and the glands in the cervix to produce barrier mucus, switching off production of fertile mucus.
Tubes	Fallopian tubes, which take eggs from the ovaries, and sperm towards the eggs. Also known as **oviducts**.
Uterus	A muscular organ, shaped like a pear, in which a baby develops and is nourished before birth. Also called the **womb**.
Vagina	Elastic muscular canal which connects the cervix to the outside of the body.

Chapter 4

A Privileged Role

"Hi, Mum," Josie called, swinging the door behind her as she and Michael came in from school. They both plonked their school bags down and came expectantly into the kitchen.

Michael opened the fridge and got himself a quick drink before calling over his shoulder, "Got football tonight".

"Can you put the kettle on for me, Josie?" Mum asked. "I'm just taking Emily round to play with Marie, and then I'll be back."

Josie was already settled with a cheesestring when Mum returned. "We were doing human reproduction in biology today," she remarked, "and d'you know they didn't even mention the **mucus cycle**!"

"Didn't they?" Mum enquired, joining Josie with a cup of tea in her hand. "Did you ask Mrs Fearnley why not?"

"That would be much too embarrassing," Josie countered, "in front of all the boys! But I did point out that the whiteboard was wrong about the life of the egg. D'you know it said that the unfertilised egg lives for several days!"

"You can see how they might think that if they don't know about the mucus and how it can keep the sperm alive," Mum said reasonably.

Josie thought before saying, "S'ppose." Then she cantered on, "And there were all sorts of other things, like saying that the **menstrual cycle**'s always 28 days long. It said that it's called menstrual because that means 'monthly' in Latin."

Mum smiled at Josie's indignant look. "Well, menstrual does mean monthly."

"Yeah, OK. But my cycle's more like every three weeks, and some of my friends are all over the place and they get really worried when people say stupid things like that. I've had to explain mucus to about five girls now!"

Josie helped herself to a banana and, with her mouth still full, went across to her bag. "Look, I'll show you. Mrs Fearnley gave us printouts."

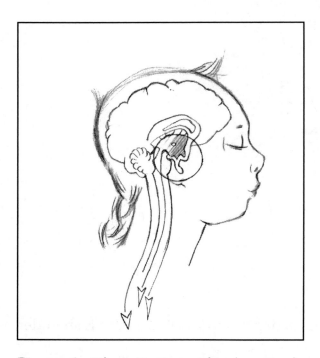

Figure 1: The pituitary gland controls many functions in the body

Mum read through the sheets to herself. "They're over-simplifying, that's the trouble with this," said Mum. "Anyway, let's go through it. You know most of this already, but we haven't done the male body, or conception. I'll get some diagrams of Dad's and the book we had last time. Can you find some paper?"

Josie obliged while Mum made herself ready. "OK," she said. "You know that, unlike the female reproductive organs, most of the male ones are outside the body. But there's one organ which is internal and is the same in both women and men. Which d'you think that is?"

Josie furrowed her brow, and then exclaimed, "I suppose it's got to be the brain."

"And the gland responsible?"

"Sorry, Mum, can't remember," replied Josie.

"It's the **pituitary gland**, and it sits at the base of the brain, somewhere in here behind the bridge of the nose," Mum said, taking up her pencil to make a sketch (figure 1 on previous page). "It's only the size of pea, and yet it produces the hormones which control all sorts of things—like blood pressure and water control in the body, as well as sexual reproduction. It works by taking messages from the brain, via this big gland called the **hypothalamus**, and responding with a range of hormones which target organs all over the body. Sometimes the targets in turn produce their own hormones, which act somewhere else."

"Like a kind of relay system?" Josie put in. "That's what happens in the ovary, isn't it, when the follicle produces the oestrogen and progesterone?"

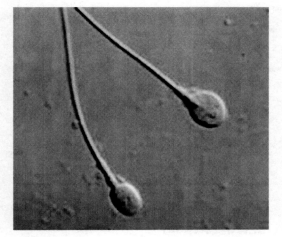

Figure 2: Two sperm

"Exactly," replied Mum. "And with a man, the sex hormone from the pituitary targets the **testicles**, or testes. The testicles are responsible for two things. First of all, they produce the male sex hormone, which is called **testosterone**—easily remembered as the name shows where it's produced. I think that your printout mentions it. Testosterone sets off and maintains the changes that happen in a boy's body at puberty, like hair and muscle growth."

"And voice breaking. Yes, we did that at school," Josie said. "We were told it influences all of the man's sexual organs and his sexual behaviour."

"The second task of the testicles is to be what I think of as a sperm factory. Inside each testis there are lots and lots of tiny tubes where the sperm are produced. But the factory only works efficiently at a constant temperature—a degree or so below that of the body. That's why the testicles hang outside (figure 3). The **scrotum** which holds them doubles up as a thermostat. If the testicles are too warm, it gets rid of the heat through its network of blood vessels, and if they're too cold, a muscle draws them up closer to the body."

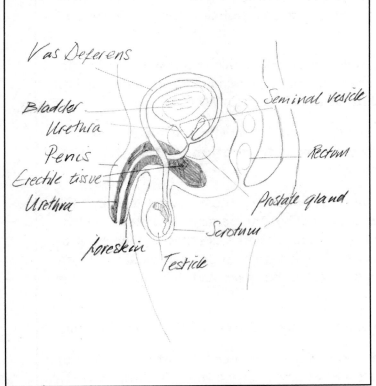

Figure 3: Male reproductive organs

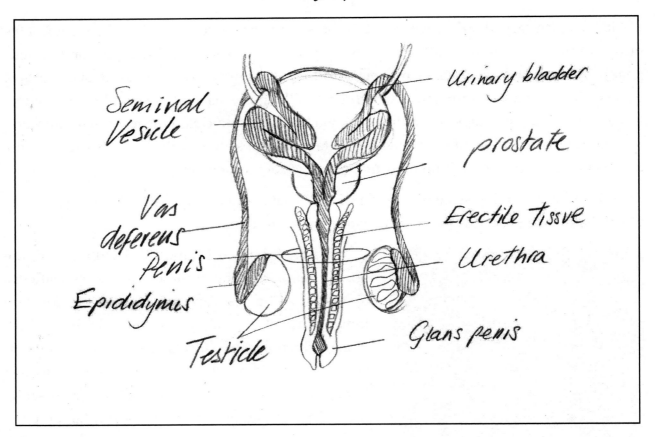

Figure 4: Sperm development and transport

The **testes** contain myriads of tiny tubes in which the sperm are produced. They need a temperature a couple of degrees or so below the body's norm of 37°C.

Behind the **testes** lies a tightly coiled tube called the **epididymis**. The immature sperm escape into this, where they are stored and matured. There is lots of room: stretched out the **epididymis** would be about 20 foot long.

During sexual arousal, contractions force the sperm into the **vas deferens**, a muscular tube which squeezes them towards the **urethra**. At this stage they still have little liquid or ability to swim.

Glands now give the sperm the fluid they need for their onward journey: the **seminal fluid**. Most of this is provided by the **seminal vesicles** and the **prostate**; small **Cowper's glands** also send into the **urethra** a slippery substance which neutralises any remaining acids. In fact **seminal fluid** is largely exactly that: fluid which feeds and protects the sperm and allows them to move freely.

The seminal fluid is **ejaculated** from the **urethra** and out of the tip of the penis.

It takes between two and three months for a sperm cell to mature. Millions of sperm are produced every day, so that at any one time there is always a fresh supply.

"That's really clever," exclaimed Josie.

"Even cleverer is the quantity of sperm the testicles produce. A healthy man can produce several billion a month."

"Several billion?" repeated Josie.

"And each of those sperm carries with it all the characteristics that a father may pass on to his child."

"But you couldn't have several billion different possibilities of being me, and every month!" Josie protested.

"Yes, you can. Every single one of those sperm has distinctive DNA, different ways of possibly being Josie. And they each contain a unique mixture of traits taken in turn from the father's two parents, which is why you can be so like Grandad.

"Of course you could have been a Joe. Men actually make the same number of male and female sperm. You'll remember about the X and Y **chromosomes** which determine the sex of a child? I suspect you did that at school?"

"The egg's 23rd chromosome is always an X, and the sperm's either an X for a girl or a Y for a boy, and when they join together at conception, to make a 46-chromosome cell, the man's X or Y determines if it's going to be a girl or boy," Josie reeled off.

"You know that pretty well, don't you? And now here's Dad's drawing of the sperm's transport system (figure 4 above). It takes between **two and three months** to mature each sperm."

Josie looked up. "Two to three months? That's ages, given there are so many of them."

"And in each **ejaculate** there are some **two to three hundred million sperm**. But the sperm, of course, are miniscule. Most of the **seminal fluid** is exactly that, a fluid for the sperm to swim through. It's added to them by various glands before ejaculation."

"If it takes various glands to produce it, it's not going to be just water," Josie thought out loud.

"No, it does several things. For a start, it neutralises the acids in the man's urethra and the woman's vagina. But it's also food. It has a really high sugar content to boost the sperm at the moment they're released."

"So that they can swim fast? It sounds very like the **cervical mucus**, doesn't it?" Josie suggested.

"When there's fertile mucus around, the two work together. But the mucus has another role which the seminal fluid doesn't have. It sieves out abnormal sperm. You see, the 'sperm factory' works at such a rate that lots of the sperm have something wrong with them. Some, for instance, end up with two tails instead of one.

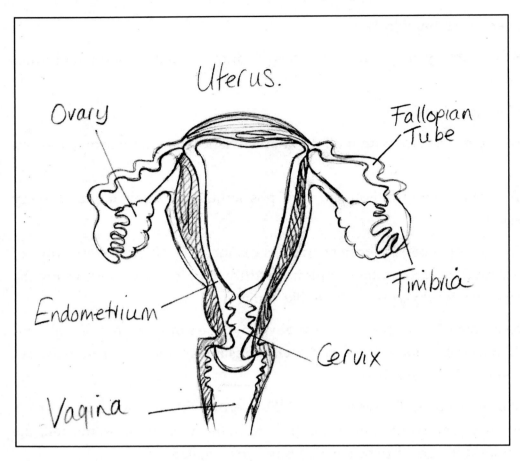

Figure 5: Female reproductive organs

"Now, let me find the diagram of the female tract," Mum said, looking through the diagrams on the table and pulling out the familiar one showing the woman's reproductive organs.

"OK, let's follow the sperm. Imagine that they're released here, at the top of the woman's vagina. If the **cervix** is shut and there's no **fertile mucus**, then that's the end of it. The sperm are quickly killed off by the vagina's acids. But before and during ovulation, the mouth of the cervix opens a little, and the fertile mucus flows out and neutralises the acids, enabling the best sperm to swim through. Once inside, a good number would appear to take a rest in the cervix or in the **endometrium** for some hours, or even longer, before setting off again in search of the egg."

"And how do they find it?" Josie asked. "I mean, it isn't exactly obvious that they should be shooting off down a tube, is it? And they might get the wrong one, too!"

"That's a good question, because the **uterus** is cavernous compared with the tubes—their internal circumference is about the size of a pinhead. Contractions do help propel the sperm but scientists have been trying to find out how they're piloted. They think they're attracted by chemical signals given off by the egg, that and possibly temperature—it's warmer in the upper part of the tube.

"So now, off the sperm go, and into the tube. By this time they're very much reduced in number, because this is a real obstacle race. Can you point out where they meet the egg?"

"You told me before," Josie replied. "It's up near the end of the tube. But at school they said it was half-way along."

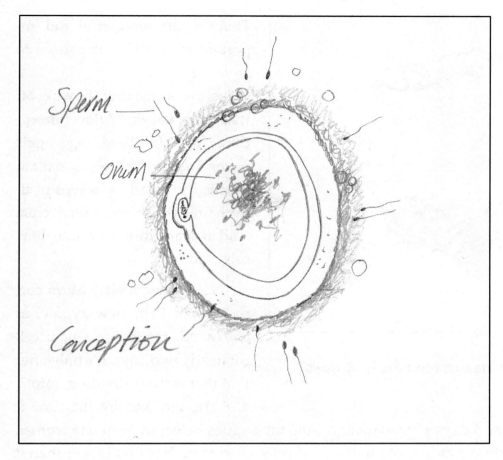

Figure 6: Conception

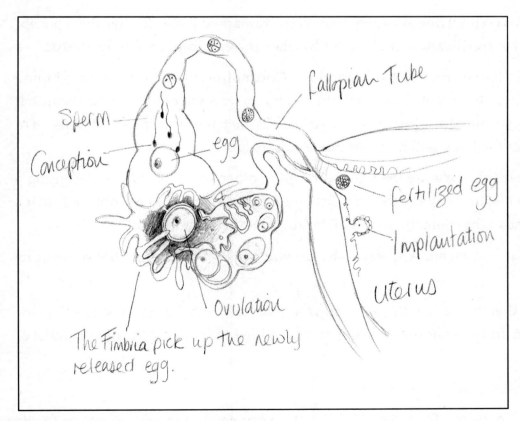

Figure 7: Conception and implantation

"Well, they've got it wrong, haven't they?" Mum replied. "The egg only lives for 12 hours or so unless it's fertilised, and it has to be pushed along by the action of the tube, so it doesn't go that far before it dies.

"Now, if one sperm penetrates the egg, something extraordinary happens. The outer membrane of the egg seals itself off so that no further sperm can enter."

Josie cupped her face in her hands. "And that was the beginning of me!" she exclaimed.

"Not quite. There's a bit of work to do before the egg and sperm combine, and it can take up to about 24 hours. It's when the two fuse that **conception** takes place, when the male and female **gametes** join together to form a new human life.

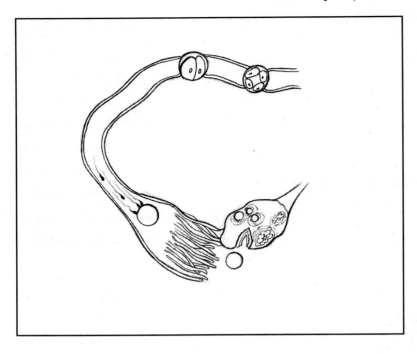

Figure 8: Zygote divides after nearly 2 days

That's a very wonderful and important moment." She paused to have a drink.

Josie was enjoying showing off what she knew: "After **conception**, the fertilised egg gets pushed along the tube, doesn't it, brushed forward by waves of tiny hairs, which are called **cilia**. And it starts dividing into new cells, too."

"Not immediately," Mum corrected her. "The new **zygote**, as it's called, remains a single cell for nearly two days. It's only after that that it starts dividing, into 2, 4, 8, 16, etc, and by the time it reaches the uterus some 6 days after ovulation—the tube varies in length from one woman to the next—it's already composed of hundreds of cells, even though it's no bigger than it was to begin with."

"That printout from school doesn't agree with you," remarked Josie. "It says that conception doesn't happen at **fertilisation** at all but only when the egg **implants** in the uterus."

"Well, that's news to me and it doesn't stand up. **Conception** means the very beginning of life. You see, the zygote's no ordinary cell. It has a life of its own, and it also has within it the full design of the human adult it will become. In fact it's imprinted with its own DNA—the same that the baby will have for the rest of its life."

She took another drink. "We use the word **concept** in other ways too, like in designing a building. The *concept* is the image of the complete building, described as a whole before it's started. It doesn't mean the moment you start digging out the foundations!"

"Hmm, I see what you mean. Anyway, why do you say that the zygote has a life of its own?"

"What I mean is that, provided it's given nutrition and the right conditions, it will grow by itself according to its own nature. Nothing more is added to it during the nine months of pregnancy."

"Like an ordinary person?" Josie asked.

"It *is* an ordinary person," Mum laughed. "Just a very small one, hidden away for its protection. For the first eight days, until it settles in the uterus, it even carries its own food with it—it only starts feeding from the mother at implantation. After that, it grows fast.

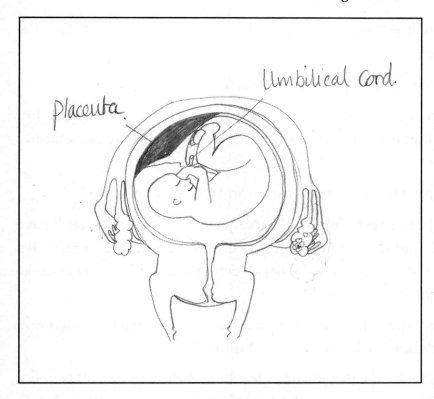

Figure 9: Baby at 16 weeks gestation

Look at this diagram of the baby, at only 16 weeks!"

"Mum," said Josie, sitting back in her chair. "I know it's all very wonderful, but sometimes it feels a bit unfair. I mean, boys can get away with having sex with girls, but girls get landed. Like Bethany becoming pregnant just now."

"It does seem a bit unfair, I know, but look at it another way," Mum reassured her. "For a start, boys have no more business than girls having intercourse before they're married. It damages them in ways which may not be so obvious, but they're there all right. Some may go on to have sexually transmitted infections (STIs), others damage their ability to settle into a good permanent relationship, let alone a lasting marriage. Unmarried fathers also often lose touch with their children and end up living with somebody else's, or with no children at all. That may not seem a big deal when you're young, but it promises a lonely old age.

"Getting back to the girl, think of yourself as the baby. You were the same person then that you are now. If I hadn't given you my protection at that time—and pregnancy and childbirth always involve an element of gift and sacrifice—you wouldn't be here, would you? Now, can you tell me who's worth more, you or me?"

"We're the same!" exclaimed Josie.

"I think so too," her mum smiled in reply. "Babies are very precious and nature has therefore seen to it that each one is shielded. The way this is done is to give ultimate responsibility for it to one designated parent. It's a privilege, not a curse, that women have been chosen for that role."

"I hadn't thought of it that way before," Josie said frowning. "You mean humans could have been like birds. They could've just laid eggs in a nest for anybody to sit on."

"With the danger, as with birds, that sometimes the eggs would be abandoned. Yes, that's exactly what I mean," Mum replied. "And the first way in which a girl is asked to protect her future young is to make sure that she doesn't get herself into a position where she may become pregnant without a husband. You see, a man's sex drive is stronger than hers. It's up to her to behave in such a way that things don't get out of hand and she doesn't lead the man into expecting sexual favours."

"Which is why you get at me about dress," Josie complained.

"And watching the time you spend alone with a boy. And various other things. But especially a girl needs to know that it's an act of real love to keep her body to herself until the man is fully committed."

"Like the egg. It makes it pretty hard for the sperm, doesn't it?" Josie remarked.

"You be like the egg, playing hard to get. You'll have many more friends, and you'll have more fun, too," Mum rejoined. "But don't think a girl is all that independent either. It's precisely because her role in motherhood will make her dependent that she needs to make sure that she has secured the right man, and for life."

"But there are lots of mothers who bring up children on their own," Josie pointed out. "Like Claire's mum, and Robert's, and in fact lots of the children in my class."

"And they can do a very good job, too. But it's hard for them. Bringing up children is a job designed for two people. The woman bears the babies and the man supports her in doing so and in looking after them. That way, they divide the work between them and share responsibility—as well as keeping each other's spirits up. Quite how they divide up the responsibilities will vary from family to family, especially as the children get older. Regardless of that, the family becomes an independent unit, able to fend for itself, with help from the wider family and community, of course.

"If the father is missing, the remaining parent has double the work and no-one to fall back on when she's tired or there are decisions to be made. For many, that's too much. The wider family may rally round, but it's still tough."

Josie grinned at her. "Don't worry, Mum, I want to get married."

"I'm sure you do, Josie," Mum replied "and I'm sure you will. It's what almost all young people want. But life isn't always that straightforward, not for everybody, and it's worth learning these things young before you get swept along in its currents. What happens in your sex life could have a much bigger impact on your future than what mark you get in an exam. So it's worth making an effort."

"What do you mean by that, Mum?" Josie asked.

"Well, on a basic level, it's easy to let your imagination carry you away."

"You mean it's wrong to think about sex?" Josie enquired. "It's kind of difficult to avoid thinking about it at all when your friends are always talking about boys and it's in about every film you see."

"No, I don't mean that. It's impossible not to think about sex sometimes—thoughts have a way of just flying into your head in any case. What you learn with practice is to cut short any thought of which you would be ashamed in public. You should also try to turn the conversation if your friends are being crude."

Mum returned to clearing up the table, "Anyway, darling, the future can bring many things. What you need to know now is that, whatever happens, Dad and I are there."

"I suppose that that too is the privilege of being a mother," replied Josie, with a grin.

Points to remember

Menstrual cycles may be called after the Latin word for month, but in practice can vary in length. Each woman establishes her own pattern.

The pituitary gland which produces the hormones controlling reproductive functions also controls a lot of other things, such as blood pressure.

Some hormones work in relay: the ones from the brain target an organ which in turn produces hormones to target somewhere else. This is what happens to the ovaries and also to the testicles.

Testosterone sets off and maintains the changes that happen to a boy at puberty, and controls the production of sperm.

The testicles are held outside the body because sperm development needs a temperature lower than the rest of the body.

The scrotum which holds the testicles acts like a thermostat, getting rid of surplus heat through its network of blood vessels, or drawing the testicles up towards the extra warmth of the body.

A healthy man can produce several billion sperm a month; there are normally 200–300 million in each ejaculate.

Each sperm has its own distinctive DNA, bearing all the traits which would be passed down to a child from the father.

It takes between two and three months for a sperm to mature.

The seminal fluid gives the sperm a medium to swim through, and protects and nourishes them.

The cervical mucus sieves out abnormal sperm, of which there are many.

Only one sperm penetrates the egg; as soon as it enters, the egg membrane seals itself off.

The zygote carries with it all the nourishment it needs until it implants in the endometrium, some eight days after conception. Apart from nutrition, nothing more is added to the baby during pregnancy: its unique DNA is complete.

GLOSSARY

Cervix — Neck of the uterus which controls the access of sperm to the female reproductive tract, especially by the alternating production of barrier and fertile mucus.

Chromosomes — Thread-like packages of DNA found in the cell nucleus. Human cells each have 46 chromosomes: 23 provided by the mother through her egg, and 23 provided by the father through his sperm. These join together into pairs. The sex of a child is determined by the man's 23rd chromosome.

Cilia — Fine hair-like projections lining the tubes, which beat in waves hundreds of times per second to transport the egg. Also present in the cervix to waft mucus.

Conception — The moment when a new human life comes into being complete with its own unique DNA.

DNA — Deoxyribonucleic acid, which is an acid from inside a cell nucleus containing the genetic instructions used in the development and functioning of living organisms.

Endometrium — Inner lining of the uterus into which the embryo implants.

Epididymis — Coiled tube attached to the testicles in which sperm are stored and mature on their way to the vas deferens.

Fallopian tube — Duct, of about 8 to 10 cm in length, which takes the egg from the ovary, and any sperm towards the egg. Conception, if it occurs, happens in the outer portion of the tube. Named after the great sixteenth-century Italian anatomist, Gabriele Fallopio.

Fertilisation — Process by which sperm fuses with the egg to form the zygote. Fertilisation can take up to a day to complete.

Gamete — Human reproductive cell (egg or sperm), taken from the Greek word for 'marriage partner'.

Hypothalamus — Highly complex organ which receives messages from all over the body and in which the body's neural and hormonal systems interact. Connects with the pituitary gland to control the latter's release of hormones.

Implantation — Attachment of embryo to the wall of the uterus.

Mucus cycle — Recurring pattern of mucus production, in which fertile mucus is visible around the time of ovulation.

Pituitary gland	Small gland in the base of the brain, which secretes the hormones responsible for controlling many of the body's functions, including sexual reproduction.
Prostate	Gland which secretes components of the seminal fluid which protect and prolong the life of the sperm.
Scrotum	Pouch made of skin, which contains the testicles.
Seminal fluid	Fluid through which sperm swim and are ejaculated from the penis.
Seminal vesicles	Pair of tube-like glands which produce the bulk of the seminal fluid, including sugary nutrients.
Sperm	Abbreviation of **spermatozoon**, the male reproductive cell or gamete which carries all the genetic contribution of the father.
Testicles	Pair of organs responsible for producing the male gametes (sperm) and the male sex hormone (testosterone). Also known as **testes**.
Testosterone	Male sex hormone, produced in the testicles, which governs the development of the male reproductive organs and other male characteristics. It also controls sperm production.
Urethra	Duct which takes either seminal fluid or urine through the penis to the outside of the body.
Vas deferens	Duct which connects with the urethra to transport sperm from the epididymis to the penis.
Zygote	Name given to the single cell formed by the union of the egg and sperm and taken from the Greek word meaning 'to be joined or yoked'. The zygote lives as a single cell for about 40 hours.

Chapter 5

It Helps to Know

Josie swung the curtain back enthusiastically as she emerged from the changing room and walked up to the mirror. "Look, Mum," she called. "Don't you think it's lovely?" She spun about expectantly, and then faltered, "Come on, Mum. What's wrong?"

"Well, I don't think that colour suits you," Mum replied, "and it's much too short."

"Oh, Mum," sighed Josie. "You're so old-fashioned! This is long compared with what everybody else is wearing!" She gave the dress a tug and added, "Anyway, I'll be wearing thick tights."

"Look, Josie. You're not having that dress. What about the red one? Go and try it on."

Josie took a last reluctant look back in the mirror before reappearing in red. "OK, Mum. What about this one? Even you can't say it's too short!"

Josie swished the skirt and straightened herself up in front of the mirror. Yes, maybe her mum was right about the colour. Red did look good on her and, even if it wasn't quite what everybody else would be wearing, it was OK. At least it was a dress.

Mum came forward and smoothed out the bodice. She stood back, wishing that she could say it was fine. Her lips pursed and she shook her head.

"Josie, I'm really sorry, but I can't have you going to a party in that! Look how low the bodice is cut!"

Josie was in a bad mood as they got into the car. They'd driven all the way to the shopping mall and spent an entire afternoon going from shop to shop, only to come away with nothing.

"Mum, what do you think I'm going to wear?" she pleaded, as she slumped into the passenger seat. "I'm not wearing that brown thing you gave me. And I can't wear the blue one—it's embarrassingly small. And I don't have anything else!"

"Josie, there's a whole month before the party, and I promise you we'll find something before then!" Mum started the car and drove out on to the dual carriageway. Then she added in a more conciliatory tone, "Do you want me to tell you why I wouldn't let you have those dresses?"[1]

Josie humphed. "I don't care," she muttered.

Mum continued regardless. "Josie, you're getting bigger now. You're becoming a young woman and if you don't know how to dress and present yourself, you're going to attract all the wrong sort of attention."

"Mum, all I want is a dress. I'm not on parade!"

"Oh yes you are! The first way anyone knows you is through your appearance. If you go along to a party showing off a lot of naked leg or a low cut bodice, that gives out a message. What people see is a body rather than a person, and boys will think about you differently."

"But they've seen my legs lots of times! In fact, every time we have a swimming lesson!" Josie protested.

"That's not the same," Mum replied. She stopped talking while she overtook, braking suddenly in front of a camera. Easing back into the slow lane, she went on: "You'll remember that we talked about puberty and the changes which take place in a girl's body? Well, the boys in your class are also gearing up, ready to father children. You'll notice that they're growing fast, and that their voices are beginning to break. This is all happening under the influence of the hormone **testosterone,** which is also stimulating their testicles to start producing sperm.

> ### The brain
>
>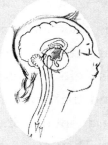
>
> **Neurons** are the primary cells of the brain. Electricity flows through them, connecting them to each other to make the brain work. Each neuron has several short projections, called **dendrites**, which receive transmissions, and one long projection, called an **axon**, to send them.
>
> The neurons are not 'wired' to each other but are linked by **neurochemicals** to connectors, called **synapses**, through which messages are passed. The synapses are created, strengthened or killed off according to the use made of them. The neurons form over 100 trillion connections with each other—more than all the Internet connections in the world.
>
> It is the impermanent nature of the synapses which allows the brain to be continually moulded throughout life according to how we behave, think and feel.

We talked about that too, didn't we? Now one of the other effects of testosterone is that it makes boys easily distracted by looking at the female body."

"So why's the swimming pool different?"

"The mind adjusts. It's normal in our society to swim with few clothes on, and it's done for a very practical reason. But can you imagine the reaction if you were to walk into class in a swimsuit? Or if you went to your party in a short nightie? It sends out a 'look at me' signal which excites boys unfairly. You see, your body isn't just preparing for future motherhood. It's also becoming attractive to boys, and the way you dress will have a big impact on how they treat you."

"How's that?" Josie asked. She was still in a bad mood, but despite herself she was beginning to listen. She had noticed that the boys looked at her differently when she tucked up her school skirt. It had flattered her and seemed like a bit of innocent fun.

"You learnt at school, didn't you, about how the body is always closely linked with the mind and spirit? The three work so closely together that you have only to think in a certain way for chemicals to be released in the brain which influence your behaviour. Boys are easily attracted by a girl's shape, much more so than girls are by a boy's."

"How d'you know?" Josie countered. She was thinking how the conversations of some of the girls round her at school dissected exactly what the boys looked like.

"I'm not saying that girls don't like a boy to be good-looking, but what really melts them is when a boy begins to show them special attention. If a boy looks at a girl lovingly, or blushes when he sees her, or makes some personal and unexpected compliment, suddenly the girl starts noticing him, even if she wasn't interested in him before. He's the same boy and may not even be one that she has particularly liked, but

> ### Neurochemicals
>
>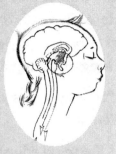
>
> Hundreds of chemicals travel about the brain's cells all of the time. Some carry messages. Others play critical roles in our thinking, desires and behaviour.
>
> **Neurochemicals** are released by the body in response to physical, mental or emotional stimuli. They are released automatically in response to these stimuli, regardless of whether the behaviour is morally good or bad.

if she isn't ready for it she can find herself disarmed."

Josie was silent and watched the cars passing. It was true that some of her friends were beginning to gossip about this boy and that one, and who was catching whose attention. She had found it really boring. They could spend so long talking about clothes and makeup and seemed to have lost other interests.

"So, Mum, are you saying that boys are looking first at girls' bodies, while girls are easily moved when a boy shows interest in them?" Josie asked.

"It's a bit more subtle than that," Mum smiled. "What I'm saying is that boys have a stronger physical instinct for sex than girls, which they need to learn to master. If you dress so that they're drawn first to your body, it isn't fair. If, on the other hand, you dress well but discreetly you give out the message 'respect me as I respect myself'.

"You see, an intimate relationship while you're still very young is likely to be unequal. The boy may want physical excitement—to show off in front of his mates—while the girl looks for affection, or to boost her self-esteem. Boys mature emotionally much more slowly than they do physically. They're just not ready for the close relationships that their physical gestures may suggest, and girls can be very hurt by that."

"So you're trying to tell me that I shouldn't have a boyfriend while I'm still at school," Josie responded accusingly.

"No, actually, I'm saying the opposite," Mum laughed, glancing down at Josie, and then quickly correcting the steering wheel. She stopped talking to concentrate, and then continued, "You should have lots of boy friends while you're a teenager. What you should be wary of is having only one. You see, if you get too involved with one boy, there will be a big temptation to progress from enjoying each other's company to enjoying a physical relationship of some sort. You will think it very innocent, and it will feel very natural and good, but what will actually be happening is that feel-good chemicals will be released in your brain, so as to draw you in and make you want a deeper and more exclusive relationship. In a girl, just being looked at appreciatively can start the process, but touch accelerates it."

> ### Dopamine
>
>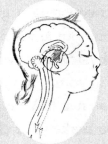
>
> This chemical has many functions in the brain. One of these is to give a feeling of intense energy and exhilaration when we pitch into demanding activities or take risks. The number of **dopamine** receptors in the reward centre of the brain declines in adolescence, so that young people need high levels of excitement. This encourages them to be bold. The **dopamine** reward creates habits of behaviour, which can be either good or bad.
>
> **Dopamine's** release can also be artificially triggered by stimulants such as alcohol, nicotine and recreational drugs. When over-stimulated, the brain learns to become resistant, needing more for the same effect and so creating addiction.
>
> Sex is one of the strongest generators of the **dopamine** reward. A couple in love are excited and confident, and prepared to take risks so that they can share their lives and live together. The sexual act then seals their love and keeps them loyal to each other. Sex indulged in lightly can trap people into wanting it for its own sake. The desire for bonding remains unsatisfied, so that there is an increasing need for excitement in sexual performance.

"Mum, how do you have a relationship without sex?" Josie asked. "I don't think you know what boys are like. They're not the same as when you and Dad were young."

"Boys aren't animals," Mum assured her, "and it's a big mistake, too, to think that human nature itself changes. The culture changes, yes, and you have to have the sense to cope with that, but people remain the same."

"You don't know the boys I know, then," Josie objected.

"The boys I know, and I do know some of your friends, want to enjoy themselves, but they also want to have good long-term relationships later on. It's just that society tells them that they can have both—that the sex they have now won't affect what happens in the future, which of course isn't true."

"So what do you expect me to do about that? Give them a lecture?" Josie grumbled.

"Boys know for themselves that there's more to sex than pleasure, physical or emotional," Mum explained. "Even the most promiscuous have ideals, and the girl they dream of marrying won't be the one who throws herself at him.

"But you have to understand things from their point of view. Their sex drive is running high and they want to hold their own before their pals. They can also be scared that they're not measuring up to the girls. So by sticking to your guns you're helping them, too. And you won't be short of friends. Take a lead and enjoy life, and you'll find others follow."

"Doesn't it help, though, I mean to have some experience, so that you know what you're doing when you meet the man you want to marry? Mightn't you otherwise marry the wrong person?" Josie asked.

> ### Oxytocin
>
>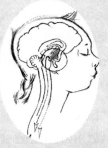
>
> This neurochemical is primarily active in a woman, and bonds her to her sexual partners, and to her babies.
>
> Any meaningful sexual contact can begin the release of **oxytocin**, even a hug. Its effects are so powerful that it can cloud a woman's judgment, and make her desire and trust a man to whom she is unsuited. When sexual relations are reserved for her husband, **oxytocin** helps a woman to overlook his faults and enjoy a stable, well-bonded marriage. The further release of **oxytocin** at childbirth and in breastfeeding cements family relationships.
>
> The breakup of a relationship is altogether more traumatic when full sexual bonding has taken place. It weakens the ability to bond satisfactorily with another man in the future.

"That sounds logical, but things don't actually work that way. You see, what you want in choosing a good partner is clear judgment. But as soon as you get involved with physical sex, you set off a host of actions in your brain which cloud the judgment you need. What's meant to happen is that you make your choice and the sexual act seals your commitment. Then a cloud of **neurochemicals** is released, binding you to your husband and dulling you to his faults.

"The chemicals are released regardless of whether or not the boy is right. So if you start the other way round, you'll still be smitten. But when you're young the structure of your two lives won't take that level of commitment. Your studies could take you apart, or you find somebody else you prefer, or one of you may have been playing around in the first place. Sooner or later the relationship falls apart, but it leaves you with a taste for physical sex so that, when you take an interest in somebody else, your relationship quickly becomes physical.

"Many people don't realise that sex bonds you in this way. It's not only the neurochemicals that cause it. Sex has such a big impact—even without full intercourse—that it creates new electrical circuits in the brain which change the way you think."

"Oh, that stuff about neurons and connectors," Josie agreed. "Yeah, we studied that at school. I still find it hard to think that there's electricity whizzing round in my brain, as though I had a whole lot of wires in my head."

"The point about brains, as I understand it," Mum continued, "is that there are no wires. The electricity leaps from one **neuron** to the next over connectors …"

"Which are called something beginning with an s … **synapses**," Josie put in.

"And these either grow or disappear according to the use you make of them. Now, when you're an adolescent, your brain is developing very very fast. There's a huge growth in the number of neurons you have, and also in the number of synapses. That's how you can take

> ### Vasopressin
>
>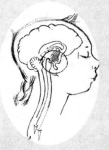
>
> This neurochemical is the male equivalent of oxytocin. It has many functions, but is specifically responsible for bonding a man to his sexual partner and to his children. Like oxytocin, it is designed to support a stable long-term relationship within marriage.
>
> Men, like women, are susceptible to losing their judgment because of premature physical bonding. Their brains are flooded with **vasopressin** each time they have sexual intercourse, producing a partial bond with each woman. Bonds thus made and lightly broken make it more difficult for a man to commit himself in marriage.
>
> Men who have many partners depend for their sexual satisfaction upon the dopamine rush. They also mould the neurological circuits of their brains into accepting multiple partners as normal.

on so many different types of activity and acquire new skills. But the growth is followed by a heavy pruning process, getting rid of pathways that you're not using, so that you can become faster and more efficient at your chosen way of life. The pathways formed in adolescence, therefore, have a big impact—which is one reason it's so easy for you youngsters to get hooked on things. A few cigarettes smoked regularly by a teenager are more addictive long term than a greater quantity smoked by an adult.

"At the same time, young people have a surge of sex hormones circulating, making you more sensitive to anything to do with sex. And then there's **dopamine**. Were you taught about dopamine?"

"The reward chemical, the one that makes you feel good when you do something exciting, like winning a race, or daring to dive off the top board?" Josie replied.

"Perhaps they said that young people have fewer dopamine **receptors** in the reward centre of the brain?" Mum asked.

Josie looked puzzled. Mum explained, "The reward centre is where the sense of satisfaction comes from. Adolescents appear to lose the number of receptors they had in childhood, and will have again when they're older. That means they need a higher level of excitement to receive the same feeling of satisfaction."

"Which is why we're sparkier than adults!" Josie rejoined.

"Well, you need to be ready for risks as you grow up. But the dopamine reward doesn't just kick in when you win a race or pass an exam," Mum continued. "It can also be short circuited by alcohol and drugs. When the blast is big, the brain responds by shutting down more receptors, and then you need a bigger dose for the same reward. It's another reason why it's particularly easy for young people to become addicted to things."

Mum paused to let Josie take this in. She then went on: "Sex is one of the strongest triggers of the dopamine reward. So if your friendship with a boy starts becoming physical, you'll enjoy the feeling of touch and want more. And if a teenager is drawn into full sexual

Prefrontal Cortex

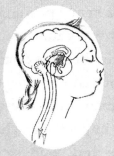

Our capacity for cognitive thought, when we make judgments, sort out priorities and control our impulses, is seated in the **prefrontal cortex**, just behind the forehead. This most sophisticated organ ultimately connects with many other areas of the brain. It only reaches maturity at the age of 23-25.

Other advanced brain organs also mature in the early 20's. These include the **amygdala** (emotion centre) and the **hippocampus** (memory centre).

By the time the **prefrontal cortex** is fully developed, most young people have already completed their studies and decided upon their adult occupations. Many have experienced their first loves, and may even have married or set up home together.

Young people have always turned to older ones for advice and we can now see that it is not only because adults have more experience: they are physically capable of better joined-up thinking.

Car insurance companies know this. They regularly charge higher premiums to the under 25's.

---

relations, she is greatly upping her chances of passing through many partners. She's also more likely to get involved in drinking and smoking, and losing grades at school."

"Mum, don't exaggerate," Josie protested. "You'll be saying next that if a girl looks at a boy she'll end up a drug addict!"

Mum laughed. "You're right. Of course I don't mean that every girl who holds hands or kisses a boy is a lost cause. Girls and boys have always done that, and many a couple has been married while still in their teens.

"The trouble today is that young people spend much longer in education, just at a time when social restraints between the sexes have been lifted. So it's difficult to keep to a chaste kiss, and once sexual relations become fully blown, you really do get trouble. You get girls suffering from what doctors call 'widow's syndrome'—deep depression, because their relationships have broken down. And boys and girls getting drunk and generally losing control, either to psych themselves up or to make up for their sense of emptiness. In any case, you get distraction from studying."

"I do agree that girls who are tied up with boys lose their interest in other things—even their friends," Josie mused reluctantly. "It makes them really quite boring."

"What's more, a history of broken relationships damages the ability to bond long-term," Mum said. "I remember hearing that when a woman has only had sexual relations with the man she marries, her marriage has an 80% likelihood of lasting, but if she adds in just one extra partner, that percentage drops to 54%."[2]

> ## Music and dance
>
>
>
> Birds and animals regularly use sound to attract a mate. Nightingales have 'singing contests' in which the male birds cut into each other with their efforts. Whale song has been found to differ in dialect from one territory to another.
>
> Human beings use all five senses in courtship, the human voice being a powerful contributor. We also use music to create a mood.
>
> **Dopamine, oxytocin, vasopressin** and other chemicals are released in the brain when we hear music. These can make us feel happy, fearful, aggressive or romantic, depending upon the type of music and our reaction to it. The music centre of the brain is actually as large as the speech centre, emphasising its importance to human culture. Strip a film or an army parade of music and they become insipid.
>
> Hearing music at its most noble can also be a deeply spiritual experience, and may variously arouse feelings of joy, reverence, awe, sadness, triumph and love. Moving to music is intrinsic to human beings, and involves the whole person in a very complete way.
>
> An important aspect of this is dance, which has always been integral to courtship; learning to dance well is a good way to attract admirers.
>
> Dance can also be abused if it is used as a cover for sexual arousal.

"Mum, what you're talking about is really serious," Josie commented, impressed.

"Sex is serious," her mum agreed. "All physical sexual contact is designed as part of a whole and is centred on the act of intercourse to which it leads. And the act of intercourse is designed to be so serious that it becomes the foundation on which **marriage** and new life are based."

"Come on, Mum. Lots of people live together without getting married!" exclaimed Josie.

"Not just lots: most people do nowadays, and some of them end up never marrying at all. Others marry and divorce, or don't marry but live with a succession of partners. It's easy to act out of impulse and find yourself sucked in, never knowing the things I'm telling you now. But sex doesn't give you a second chance. Once you've lost your virginity, it's gone. You might become **pregnant**. You might pick up a **sexually transmitted disease**—I'll talk to you about this another time. But even if you don't conceive or pick up a disease you'll be changing your body and with it your life chances. Marriage isn't just a social invention. It's written into what we are designed for. Can you think why?"

"It's something we dream about?" Josie proposed.

"Yes, that's a good point," her mum agreed. "It's natural to dream about getting married. But apart from the romance and the beautiful dress and everything, what are you actually dreaming about?"

"Being together forever?" Josie suggested.

"Like in a love song? But that's actually it: being together with one person, which means being faithful to that one person, forever, which means until one of you dies. Now look at the act of intercourse. The body is saying the same thing: 'I love you so much that I want to share everything I have with you, even to my seed'. The seed means either the man's sperm or the woman's egg. There's nothing more solemn or more exclusive and long lasting than sharing in giving life to a new human being. So whether or not a child is actually conceived, the giving of the seed by the man to the woman is a sign of lasting and exclusive love. It belongs to marriage."

"And the brain is saying the same thing, when it sets up all those synapses and everything," agreed Josie.

"So what do you think happens when you have a sexual relationship without that commitment?" her mum asked.

"It's not honest," Josie considered, "because your body is still giving everything, but your mind and spirit aren't doing the same."

"That's very well put," Mum replied. "I may say, at the time, the actions of the body can make you feel so good and so close that they can convince you that you are giving everything. Sex tempts everybody, unlike drink or drugs."

"From what you say, that's got to be," Josie replied. "You only become addicted to outside things if you experiment with them in the first place. But sex is inside you, isn't it?"

"Yes, it's part of you," Mum agreed. "That's why you have to use your brain to make sure that you don't cheat on yourself, too!"

She paused as she drew up at a set of traffic lights. They took a long time to change and she stretched herself before relaxing back into her seat. "There's another danger," she remarked. "You don't have to be with someone else to take a selfish pleasure from sex. You can get physical pleasure by applying pressure to your sexual organs, which is called **masturbation.** This again is a cheat, because you're using your body for yourself rather than reserving it as a gift for your husband. If you are tempted, just get up and do something different, and don't let yourself get carried away with daydreaming."

"But Mum, what's the harm in masturbation?" Josie asked.

"Well, for a start the brain is making those emotional ties when there's no man to attach itself to. So it's deceived, and you can get into bad habits which are difficult to break. And then there's the mental imagery which comes with physical stimulus. Your body, mind and spirit are never more inter-connected than when it comes to sex. You can think yourself into situations which would make you blush if they were spoken out loud."

They were approaching home now, and their conversation became more immediate. They had been out for much longer than intended and Mum was working out what food there was in the house to cook at speed. Josie in the meantime returned to the subject of her dress. Mum had a bright idea.

"Look, darling, I'll take you to the boutique in the High Street. They're having a sale, and we might just find something in there. I know they have small sizes."

Josie was chuffed. Her mother occasionally went into the boutique herself, but had never before suggested taking Josie there. "We won't tell Daddy, OK? We'll give him a surprise and get you something really nice." She smiled at Josie's flushed face and added, "You see, when you dress well, you give pleasure to lots of people. And life isn't only about finding a sexual partner; it's about making friends and getting on with everybody round you, male or female. You only have to find one husband, but you'll be mixing with other people for the rest of your life."

Points to remember

The first impression people have of you is through your appearance. The way you present yourself automatically gives off a message.

Different clothes suit different occasions and different people. We show respect for ourselves, and for others, by making an effort to dress appropriately.

Boys are easily attracted by a girl's appearance and can be unfairly aroused if girls dress immodestly.

Girls tend to look first for affection in a relationship, and are readily disarmed by boys taking an interest in them. They are prone to mistaking physical intimacy for love.

Boys, like girls, have high ideals and want to marry, but they mature more slowly and are rarely ready for commitment in adolescence. Teenage sexual liaisons easily become occasions in which boys show off to male peers.

Neurochemicals prompt sexual attraction and are released at intercourse in such numbers that they impact upon the connections in our brains, changing the way we think and behave. This is designed to bond us permanently to sexual partners.

The brain has a surge of development in adolescence, multiplying in neurons and synapses. These are then heavily pruned according to use, making the brain especially malleable in youth.

Sexual experience clouds our judgment. Liaisons which have been consummated in the teenage years rarely survive into adulthood and cause a girl much heartache when they break down.

Breaking a sexual bond damages the ability to bond in the future.

Young people are generally more prone to addictions than adults, because of the way the dopamine reward works, and because the pathways of their brains are still forming.

Nowadays they face added difficulties since education lasts longer and new cultures have removed many social restraints.

GLOSSARY

Amygdala	Almond-shaped groups of nuclei which perform a primary role in processing and storing emotional reactions.
Axon	Long projection attached to the neuron for transmitting messages.
Brain scan	Images the brain's activity by measuring its points of energy release and blood flow. Developed in the 1990s, it is allowing scientists to study the brain's pathways.
Dendrites	Short projections attached to the neuron for receiving transmissions.
Dopamine	Neurochemical with many functions. These include releasing from the reward centre in the brain feelings of intense satisfaction and pleasure in response to exciting or demanding behaviour. The **dopamine reward** plays an important role in many aspects of human behaviour, including the sexual. Dopamine release can be over-stimulated with the use of drugs, alcohol and casual sex, leading to addictions.
Hippocampus	Major component of the brain's limbic system which consolidates short- and long-term memory and spatial navigation.
Oxytocin	Neurochemical, especially active in a woman, which is responsible for feelings of bonding towards a man and towards her child. Oxytocin is released in large quantities during the sexual act, at childbirth and in breastfeeding.
Prefrontal cortex	Area of the brain responsible for executive functions, such as defining goals, setting priorities, assessing consequences and controlling impulses. It only matures at age 23-25.
Marriage	Is only legally ratified at intercourse, hence the alternative term 'marriage act'.
Masturbation	Exciting one's own sexual responses as a solo act.
Neurochemical	Chemical which bathes the brain cells and carries messages across the synapses from one cell to the next.
Neurons	Primary cells of the brain through which electricity flows to make the brain work. By the end of adolescence, a person's brain will contain more than 10 billion neurons.
Pregnant	Girls who are sexually active have a much higher rate of unexpected pregnancy than more mature women even when they use contraception. (See also chapter 10.)

Receptors Molecules attached to target organs which 'catch' hormones and other substances as they travel through the blood stream. Each receptor is hormone-specific, though it may work with more than one. Once a hormone is bound to a receptor, it triggers a cascade of reactions within the cell affecting the cell's function.

Sexual diseases Diseases and infections acquired through sexual contact with a carrier. They are usually referred to as STDs (sexually transmitted diseases), or STIs (sexually transmitted infections), and are particularly catching among young people whose defence mechanisms are not yet fully mature.

Synapses Connectors which help carry electrical messages from one neuron to the next. The brain is not hardwired: the gap between the neurons and the synapses is bridged by neurochemicals. Synapses form and disintegrate according to use, making the brain continually adaptable, but especially during adolescence.

Testosterone Male sex hormone which governs the development of the male reproductive organs and controls the male sex drive.

Vasopressin Neurochemical which functions in a man much as oxytocin does in a woman, bonding him to his mate and to his children.

Chapter 6

Michael and Dad Go Fishing

MICHAEL AND HIS DAD had had a good afternoon. They had caught several fish and Dad had been showing Michael how to gut them. "Look, Dad," the boy called. "This fish has got lots of little eggs inside it!"

"So it has," Dad replied, pulling open the fish. "It must have been female. But it's not going to have any more babies now," he added, cleaning it out and throwing it into his sack. He looked across at his son, already absorbed in the next fish.

"This one's got some also," remarked Michael. "But I can't see any tails, so they must be very young."

Dad looked puzzled at first, then laughed. "They're not like those frogs' eggs that you were looking at last spring," he said. "Those had already been fertilised by the male frog when they were laid. These haven't been fertilised, so they're not yet growing into little fish."

He's really growing up, he thought to himself. *In fact, he'll be a teenager soon, and it's about time I spoke to him.*

"Do you know how eggs are fertilised?" he said out loud, casually, as he returned to the tackle he was sorting. "Or how a bull covers a cow before a calf is born?"

"Of course I do," replied Michael scornfully. "We've been learning all about it in science."

"You have, have you?" Dad asked. "So how much d'you also know about how you were made?" he said, hoping that he sounded natural.

Here goes, thought Michael, *and I've known this for ages.* Their science lesson had continued out in the playground where some of his schoolmates had been showing off to each other. Out loud he said, "Everyone knows that one of the man's sperm has to join the woman's egg to make a baby."

"So you know about how a man's sperm joins the woman's egg to begin a new life?" Dad repeated.

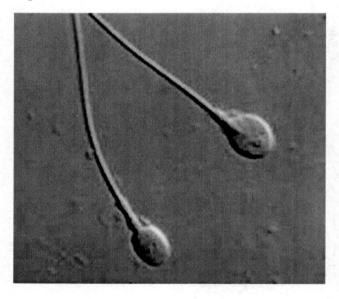

Figure 1: Sperm under a microscope. The head contains the DNA and the tail enables it to swim

Michael shrugged. "Well, sort of," he replied vaguely.

"I see," Dad replied. "I think it's about time you knew a bit more. We'll start at the beginning, shall we, and talk about the man's sperm. I don't suppose you've ever thought about how they're produced in a man's body? How many there'll be, or how long they take to mature?"

Michael looked puzzled. "Don't they just swoosh out?"

"They can't swoosh anywhere if they aren't there, waiting for action," Dad replied laughing.

He propped up his rod and caught Michael's eye. "You already know quite a bit about your own anatomy, and mine too for that matter, but some of it you probably don't know.

"Starting about now, and going on for the next couple of years, you're going to begin the process of turning from a boy into a man. That's a big step. Many of the signs are obvious—your body starts growing very fast (and your appetite with it), your voice will break and become much deeper, your muscles will increase in size, and you'll grow hair on your face and other parts of your body. All these things happen at different times, and in a different order, depending on the boy. But what would you say is the critical difference between being a boy and becoming a man?"

Michael looked completely blank. Then he blushed a bit and said awkwardly, "You can start getting girlfriends and marrying, and all that sort of stuff."

"That's all part of the answer," Dad laughed, "though you're a bit young to be thinking of getting married while you're still at school! No, I'm talking about something more general, something that happens to every boy at **puberty**."

Michael pursed his lips and thought. "I dunno," he said at last.

"If you look it up in the dictionary, you'll find that the word 'puberty' means 'becoming able to procreate children', or becoming fertile." Michael still looked blank, so Dad continued, "It means that your body will start producing the seed which enables you to reproduce. For a boy, that's becoming able to father a child. Most of the other changes in your body relate to that one change.

"A man goes on being able to father children for the rest of his life. His sperm decrease in number and quality but he's still able to do it. A woman's different. She's much less fertile after the age of 40, and stops being so in her late 40s or early 50s—that's nature's way of making sure that mothers of babies have the stamina they need—but a man, in ordinary circumstances, can father a child till the day he dies."

"Yeah, on the news, d'you remember, that man fathered a child at the age of 90! I don't call that ordinary. I call that weird!" exclaimed Michael.

"Just because it can be done doesn't mean it should be done. Same with having close contact with a woman. The fact that your body is gearing up for fatherhood doesn't mean to say you're ready for it. OK, next question. What do you think is the principal sexual organ?"

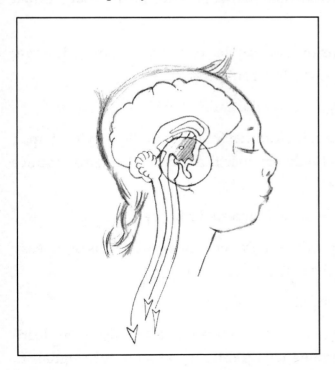

Figure 2: The pituitary gland in the brain controls many functions, including the male and female sexual organs

Michael thought for a minute, then grinned. "It's got to be the brain."

His dad looked surprised. "Well done. I certainly didn't know that at your age. Can you tell me why?"

"Yes, actually," Michael replied. "We've been learning at school about how our brains are wired up. Men's and women's are different, and that affects all sorts of things, the way we behave, our feelings and all that kind of stuff. It always seems to be the brain that starts everything."

"Good lad, you're spot on there. Everything to do with creating babies is controlled by the brain. But not everything happens in the brain—the brain's job is to give out instructions to other organs, and it does so using special chemicals."

"Oh?" queried Michael.

"They're called **hormones,** but you can think of them as chemical messengers. They travel through the blood stream from one organ to another. The receiving organ has to have **receptors** to grab the hormones as they pass. Come here and I'll show you what happens to produce a man's sperm." He rummaged in his knapsack for a notebook and pencil.

"OK," he said, settling down. "So we have hormones coming down from the brain with their instructions. And they target these organs down here, which are the **male reproductive organs.**" He began drawing (see figure 3).

"The shape of all of this is going to be very familiar to you, because most of the male's reproductive organs are held outside the body. I'm putting in the **foreskin** which protects the end of the **penis**, though many boys have it removed soon after birth. That's called **circumcision.**

"This is where the sperm are first made, inside the **testicles** or testes. I'm only drawing one, but you actually have two and they live inside this protective sack."

Michael watched him label the sketch. "D'you mean the balls, Dad?"

"Yes, Michael. Testicles are also called your balls, but they're not round—they're shaped a bit like fat rugby balls, in fact. And they don't only produce the sperm. They also produce a very important hormone called **testosterone**."

"So it isn't just the brain which produces hormones?" queried Michael.

"No, the brain can prompt other organs to produce them too. The name of this one's easy to remember: test—osterone, or 'I come from the testes'."

"Yeah, I've heard of testosterone," said Michael.

"It's important, because, among other things, it controls male characteristics. Your testicles grow in puberty, and so does the testosterone they produce, and that in turn signals to the body that it's time to become a man."

"Like having a deep voice and all that stuff?"

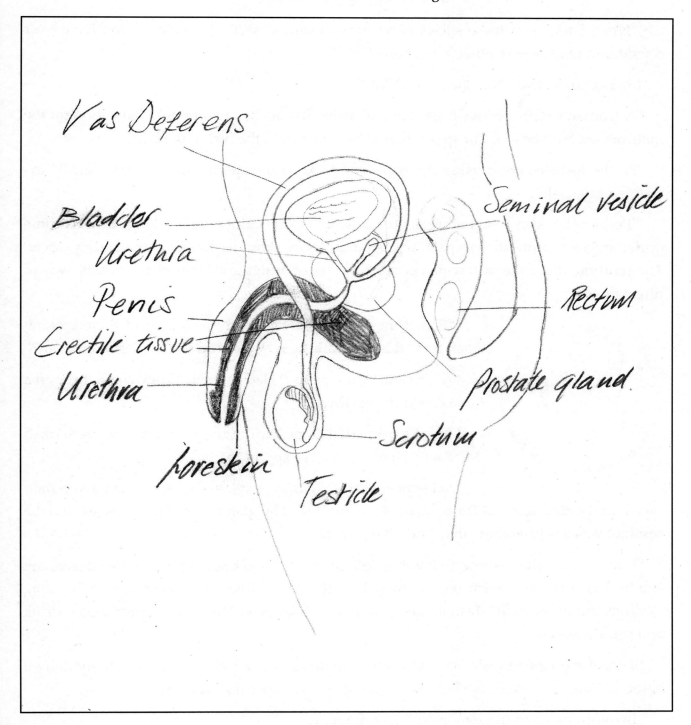

Figure 3: Male reproductive organs

"Yes, and hair growth and muscle development. At the same time, your hormones affect the brain itself. You become more aware of yourself and want to assert your own personality. That's why teenagers can be rebellious and selfish: they're establishing their adult place in the world and in the family. Young people need to learn how to channel their strength and their energy, so that they focus on what should be rebelled against—like injustice—and not against what is good, like the family who love you.

"Now, I told you the testicles hang down inside a sack. I wonder if you have ever considered why they're outside the body?"

Michael frowned. "No," he said. "Why?"

"A woman's reproductive organs are all inside her body, so you might have thought the male ones would be too. But sperm like to be cooler than the rest of the body."

"So the testicles are outside the body just to keep the sperm cool? That's weird!" exclaimed Michael.

"Perhaps it's also designed to make us aware of our vulnerability! If the testicles come under any sort of physical impact, we certainly know about it. But again the body's clever. The **scrotum,** that's the sack which protects the testes, is designed to move out of the way of direct blows."

"Sometimes," Michael said with feeling. And then, looking back at the diagram he asked, "Where do all the sperm come out?"

"At the end of the penis," Dad replied. "They come through a tube called the **urethra**."

"But it's urine that comes out of the penis, surely," remarked Michael. "Does urine have sperm in it?"

Dad smiled at this. "I didn't explain that sperm are given their own liquid called **seminal fluid**. Most of it is produced by glands called the **prostate** and the **seminal vesicles** just before they enter the urethra.

"If the urine contained sperm it would kill them off, it's so acid. What I've been describing is a bit like a points system on a railway line: the penis either has sperm or urine passing through, but never both." Dad looked again in his knapsack. This time he produced a drink and some biscuits.

Michael munched in silence. There was a lot to take in which he hadn't actually known before. "Dad," he asked, "so how long does it take for sperm to develop?"

"Between two and three months," was the reply.

"So men can only fertilise a woman every few months?" Michael asked.

Dad shook his head. "No, we can make love when we want, within reason. There's a constant supply of sperm coming from the testicles. And, as I said, most of the fluid is from the prostate and the seminal vesicles, which the body makes quickly because it's much less complex than sperm."

Michael was impressed.

Then Dad added, "Do you remember you told me you were learning about **chromosomes** at school?"

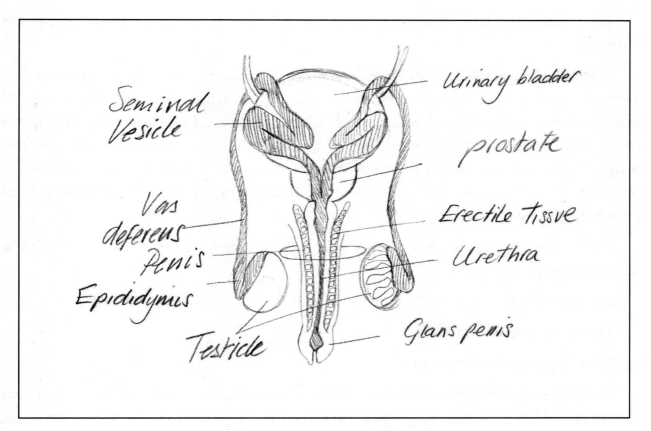

Figure 4: Sperm development and transport

The **testes** contain myriads of tiny tubes in which the sperm are produced. They need a temperature a couple of degrees or so below the body's norm of 37°C.

The **scrotum** holds the **testes** and also insulates them: if it is too cold, the **scrotum** draws them closer to the body; if it is too warm, it loses surplus heat through a network of blood vessels.

Behind the **testes** lies a tightly coiled tube called the **epididymis**. The immature sperm escape into this, where they are stored and matured. There is lots of room: stretched out the **epididymis** would be about 20 foot long.

During sexual arousal, contractions force the sperm into the **vas deferens**, a muscular tube which squeezes them towards the **urethra**. At this stage they still have little liquid or ability to swim.

Glands now give the sperm the fluid they need for their onward journey: the **seminal fluid**. Most of this is provided by the **seminal vesicles** and the **prostate**; small **Cowper's glands** also send into the **urethra** a slippery substance which neutralises any remaining acids. In fact **seminal fluid** is largely exactly that: fluid which feeds and protects the sperm and allows them to move freely.

The seminal fluid is **ejaculated** from the **urethra** and out of the tip of the penis.

It takes between two and three months for a sperm cell to mature. Millions of sperm are produced every day, so that at any one time there is always a fresh supply.

"You mean that cells have 46 chromosomes which control the way they are?" suggested Michael.

"That's right," said Dad. "Well, sperm cells and the woman's egg cells are different from every other sort of cell. When they mature, ready for action, they each have only 23 chromosomes. Now, when we began this conversation, you told me that one of the man's sperm had to join with the woman's egg to make a baby. You can see how cleverly this happens. The man and the woman each provide cells with 23 chromosomes so that, when they fuse, the new baby has a complete set of 46. That's how children inherit characteristics from both parents.

"Tell me something else. How many sperm do you think a fertile man sends out to fertilise an egg? Have a guess."

Michael paused. "A million?" he ventured.

"Try again," replied Dad.

Michael shrugged, "I've no idea."

"The answer is **two to three hundred million**. In fact, a healthy man can produce **several billion** sperm every month. Staggering, isn't it, when you think that one sperm carries all the characteristics a father passes on to his child."

Dad got up, stretched and started packing away the tackle. "Mum's going to wonder what's happened to us," he remarked, giving Michael a line to untwine. Michael deftly loosened the knot and they were soon ready to make their way back to the car.

"Dad," Michael asked, as they walked along. "Is a wet dream something to do with sex? I heard them talking in the playground the other day, and I didn't want to let on that I didn't understand."

"Good point. I should have mentioned that," Dad replied. "When you grow into puberty, you may find that occasionally, while you're asleep, you'll have an **ejaculation** of seminal fluid. It just happens in the night. That's called a **nocturnal emission** or a wet dream. It's basically your body gearing up for intercourse and making sure that everything's in running order."

"Is that how spare sperm are removed from the body, then?" Michael asked.

"If all those billions of sperm were to be removed like that, you'd be having wet dreams on a regular basis! Yes, some will be removed like that, but most are just reabsorbed back into the system again. In fact, some men never have wet dreams at all. It's nothing to worry about, one way or the other. What you don't want to do, though, is to stimulate your sexual organs yourself."

They had reached the car and, as they got in, Dad said, "Anyway, now you know you're not just one in a million, but one in several billion. There are lots of other things to tell you—we've only just started really—but they'll wait for another day. What do you say to some fresh fish for supper?"

Points to remember

Puberty describes the stage in an adolescent's life when the body's reproductive organs begin to function.

Other essential changes relate to that one change. For a boy, this includes fast growth in height and muscle, the voice breaking, and hair appearing on the face and other body parts.

The principal sexual organ in both men and women is the brain which controls all the other reproductive organs as well as the way we think and feel. This is different in men and women.

The male reproductive organs are held outside the body. The testicles produce millions of sperm every day at a temperature cooler than the rest of the body. They are insulated by the scrotum in which they hang.

The brain controls the reproductive function through 'chemical messengers' called hormones, which travel through the blood stream. The receiving organs have receptors, which take hold of the hormones as they pass.

The hormone testosterone is made in the testicles. Testosterone controls sperm production and the development of male characteristics. It in turn acts on the brain, affecting the way a man thinks and feels.

It takes between two and three months for sperm to mature. There are about two to three hundred million of them in each ejaculate. The rest of the seminal fluid is largely composed of a protective fluid made in the prostate and seminal vesicles.

A healthy man keeps the capacity to father a child for the rest of his life. A healthy woman becomes less fertile as she grows older, and ceases to be so in her late 40's or early 50's.

At maturity, the sperm and egg cells have 23 chromosomes each, ready to fuse into one new cell of 46 chromosomes at conception.

Boys may experience wet dreams or nocturnal emissions when seminal fluid is ejected spontaneously during the night. This is the body making sure that everything is in running order and is nothing to worry about.

GLOSSARY

Circumcision — Cutting off the foreskin covering the tip of the penis. The practice dates from ancient civilisations and has religious as well as medical connotations. Jewish males are circumcised in a ceremony of initiation into the faith; circumcision is also widely practised by Muslims. Some doctors use circumcision against urinary infections and some sexually transmitted diseases.

Chromosomes — Thread-like packages of DNA found in the cell nucleus. Human cells each have 46 chromosomes: 23 provided by the mother through her egg, and 23 provided by the father through his sperm. These join together into pairs. The sex of a child is determined by the man's 23rd chromosome.

DNA — Deoxyribonucleic acid. DNA is present in the nucleus, or control centre, of each of the cells from which living beings are made. Inside the DNA there are tiny **genes** which contain instructions for how cells are to grow and behave. Each person's DNA is unique and every cell of our body is marked with that unique DNA, except for our eggs or sperm which, remarkably, each have their own DNA.

Ejaculation — Ejection of seminal fluid from the penis.

Epididymis — Tightly coiled tube attached to the testicles. Sperm are stored and matured in this for a couple of months or so before they are ready to pass into the vas deferens.

Foreskin — Fold of skin covering the end of the penis.

Hormones — Chemicals which circulate through the blood-stream, giving instructions from one organ to the next. Hormones can originate in the brain, or in other organs such as the ovary. They often trigger each other in relay.

Male reproductive organs — Network of organs which enables a man to father children. The reproductive organs produce, store and transport sperm, and deliver them to the female at intercourse.

Nocturnal emission — Commonly known as a **wet dream**. Spontaneous ejaculation of seminal fluid as part of a sleeping dream.

Penis — Male sex organ which releases sperm into the vagina.

Prostate — Gland which secretes components of the seminal fluid which protect and prolong the life of the sperm.

Puberty — Phase of adolescent life when children's bodies mature into adulthood with the capability of sexual reproduction.

Receptors	Molecules attached to target organs which 'catch' the hormones and other substances as they travel through the blood-stream. Each receptor is hormone-specific, though it may work with more than one. Once a hormone is bound to a receptor, it triggers a cascade of reactions within the cell, affecting the cell's function. Hormones can both start a process and stop it.
Scrotum	Pouch made of skin which contains the testicles. It insulates and protects them.
Seminal fluid	Fluid through which sperm is ejaculated from the penis at intercourse.
Seminal vesicles	Pair of tube-like glands which produce the bulk of the seminal fluid, including sugary nutrients.
Testicles	Pair of organs responsible for producing the male gametes (sperm) and the male sex hormone (testosterone). Also known as **testes**.
Testosterone	Male sex hormone, produced in the testicles, which governs the development of the male reproductive organs and other male characteristics. It also controls sperm production.
Urethra	Duct which takes either seminal fluid or urine through the penis to the outside of the body.
Vas deferens	Duct which connects with the urethra to transport sperm from the epididymis to the penis.

Chapter 7

Michael Shows Off his Knowledge

"It's not fair," complained Michael, stomping to the kitchen. "It's always my turn to do the washing-up."

"Our turn, you mean," said Dad, following him with a load of plates. "We'll have this done in minutes, you'll see."

He put on a large apron and turned on the tap. "Anyway," he remarked, "it gives me a chance to ask you how school's going. How was that maths test?"

Michael reluctantly took a tea-towel and picked up a glass.

"Fine. I was well above pass mark." He paused and then added, "Hey Dad, did I tell you? Mrs Simmons, our maths teacher, she's having a baby."

They continued talking for a bit and then Dad said, "You mentioned Mrs Simmons and her baby just now. I think it's time you learnt about babies and how they come to be. Look," he put some last dishes into the sink, "just dry these few things and the rest can drain. Mum's got a book upstairs. I'll see how much you know."

He made himself a mug of coffee and pulled up some chairs. "Remember when we talked about the development of sperm? Now we can fill in on the woman's body too."

"Yeah, we've done that in biology," Michael said, but he tucked himself on to a chair all the same and got some paper.

"In that case, you can teach me," smiled Dad.

"OK," said Michael. "I can even draw it for you." He took the paper and began sketching, while Dad looked on, impressed.

"You have to imagine this is inside the woman's body," he explained. "The big thing in the middle is the **uterus** where babies develop, and on each side there's an **ovary** where the **eggs** start out, and above them these long wiggly things are the **fallopian tubes**.

"I'm going to start with an egg in one of the ovaries."

Dad interrupted, smiling, "But you haven't told me what this egg is for."

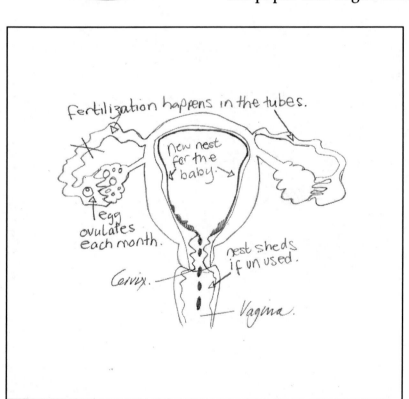

Figure 1: **Michael's summary of menstrual cycle**

"It's the woman's reproductive cell, like the man has **sperm**. Only it's different with a woman, because all the eggs are inside her when she's born—in fact she has about half a million. You can see a few here," he added, drawing in some circles of different sizes in one of the ovaries.

"Right. Every month, an egg matures. It gets bigger and bigger until it breaks out of the ovary, and that's called **ovulation**.

"The egg then gets picked up by the **fimbria**, which are feathery fingers attached to the **fallopian tubes**. And if it's fertilised, that happens here in the tube." He drew a large **X**. "But most of the time it isn't fertilised, and it dies in 2-3 days.

"After another 10 days or so the **uterus** sheds its lining, and the woman bleeds. That's called **menstruation** because it happens every month. Oh, I forgot to say that the **lining of the uterus** is important because it acts like a nest. Each month a fresh lining builds up in case there's a baby, because that's where it **implants** and grows until it's ready to be born."

"You've learnt a lot, haven't you? Let's begin again with all those basics still in mind. Every cycle, a few eggs are chosen for development. Can you tell me what starts the process off?"

"It's a hormone, isn't it? From the brain, I mean."

"Good lad, you've remembered that the brain's in overall charge. So a hormone from the brain tells the ovary that it's time to select some eggs. These don't mature on their own, do they, but they grow inside protective cyst-like structures called **follicles**. And one of these grows incredibly fast until it takes over from the others—which is why it's called the **dominant follicle**."

Inside Mum's book Dad had found the drawings Mum had done for Josie. He looked through them now to see what might be there.

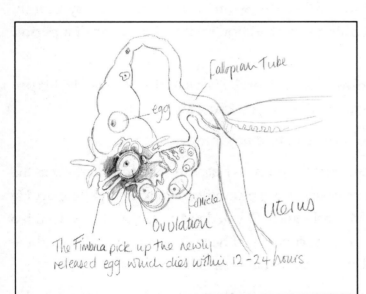

Figure 2: Ovulation and death of egg

"Did you know that the egg inside that follicle becomes the biggest cell in the body?" Michael remarked as he waited for Dad. "By the time it's released you can even see it without a microscope—it's about the size of a full-stop."

"Well, here you see it all," said Dad producing a diagram of ovulation. "You have to imagine that that large egg of yours also grows round itself a follicle the size of a walnut. In fact, the dominant follicle's so big it literally breaks open the wall of the ovary before surrendering the egg."

Michael looked at the diagram with interest.

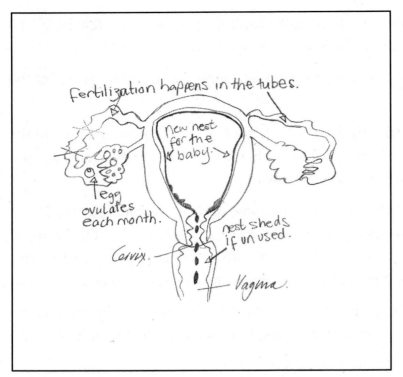

Figure 3: Michael's amended drawing marking the site of conception in the outer tube

"And, what's more, Mum's written in that the egg, unless it's fertilised, dies not within days but within 12–24 hours. What do you think that means for your diagram?"

Michael frowned.

"Well," said Dad, "it takes about six days for the egg to reach the uterus. You've marked your X for fertilisation about half-way along—which would be right if the egg really lived 2-3 days as they told you at school. But if it only lives for 12-24 hours, it would be dead by then, wouldn't it?"

"Which means fertilisation has to happen within hours of ovulation?" queried Michael.

"Exactly. And it has to happen up towards the ovary end of the tube. Try again with your cross," he said.

"There's something else you ought to know. We say women menstruate every month, but in practice the timing can vary a lot. Calling menstruation by its other name of a **period** perhaps makes that clearer."

Dad looked at Michael's diagram again. "How about the **cervix**?" he asked, pointing with his pen to the neck which joined the uterus to the vagina. "What do you know about that?"

Michael shrugged and said, "Dunno. We weren't told anything at school."

"Weren't you?" Dad exclaimed. "But **mucus** from the cervix is just as important for fertility as the egg and the sperm." *No wonder they got the life of the egg wrong,* he muttered to himself, getting up. He helped himself to some more coffee, offering Michael a juice. "If you don't understand how the three organs work together, the ovary, the uterus and the cervix, you haven't begun to understand the woman's reproductive cycle. It's complicated, but I'll try to make it as simple as I can."

Returning to the table, he took the book. It opened at a photograph of **fertile mucus** under a microscope (figure 4). He then found the large diagram he was looking for, which he smoothed open (see figure 5 overleaf).

"This should help explain it. You can see the brain at the top with its two-way arrows to the ovaries. That shows that there's on-going interaction between the two.

Figure 4: Fertile cervical mucus under a microscope

"Then, below, you have the two ovaries, the uterus, and the cervix. I should say that anything I'm now going to describe could be happening in either ovary—they work the same way but in different cycles. And then at the bottom you have a full picture of the uterus and other organs.

"But can you see that the diagram also divides up between left and right?"

"'*Before ovulation*' and '*After ovulation*'", Michael read out.

"That's right," Dad said. "So, as well as the period, which is represented at the bottom of the page, there's another big event in each cycle, and that's ovulation. What happens before it and what comes after has major significance.

"Looking at the left-hand side, '*Before ovulation*', you can see the follicle developing in the ovary. What else is it doing?"

"It's producing high levels of **oestrogen**," Michael read out.

"OK, so the follicle is producing a hormone called oestrogen and, if you follow the arrows, you'll see that the oestrogen acts on the uterus, and on the cervix."

"I know about the uterus," Michael said. "It'll be building up the lining in case there's a baby."

"So it is. What the oestrogen's doing in the cervix is really clever. Just for a few days it produces mucus like in that picture I've just showed you. You'll see it has lots of channels running through it for the sperm to swim through. On they go, through the **cervical canal**, and into the uterus. From there they can seek out the egg in its fallopian tube."

"So the mucus appears just when the egg's around?" Michael asked.

"And beforehand. Round the time of ovulation," Dad replied, "which is why it's called fertile mucus. Now look on the '*After ovulation*' side. You can see that the follicle has collapsed, because its egg has gone. But the collapsed follicle, under the new name **corpus luteum**, continues to be important. It produces the hormone **progesterone**. Can you follow the arrows and see what it does?"

Michael looked. "It says it makes the **endometrium** secrete nutritious juices, presumably to make the uterus even readier for a baby. And the second arrow says it blocks the production of fertile mucus in the cervix."

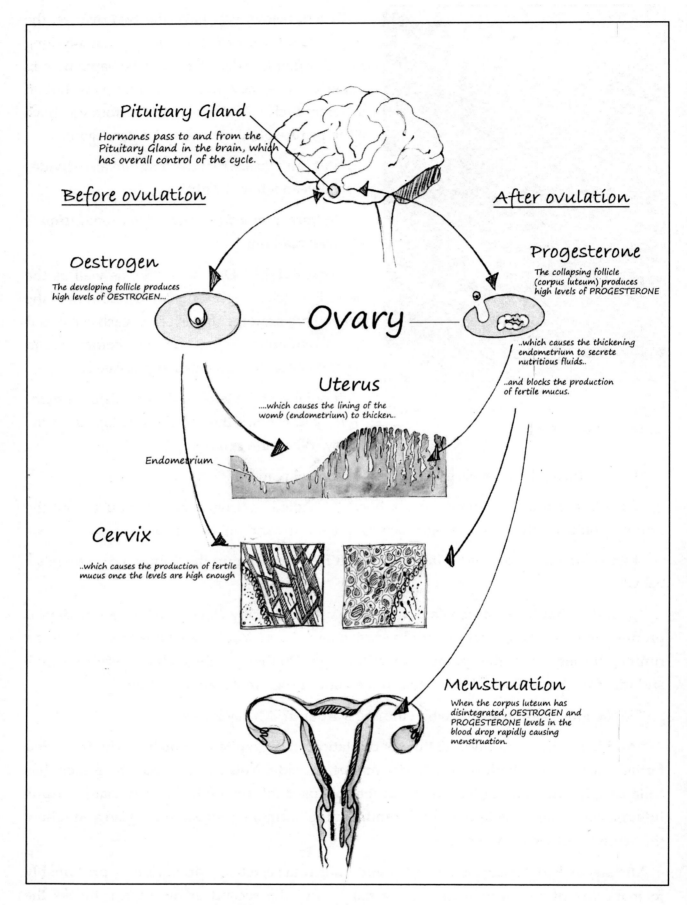

Figure 5: The menstrual cycle

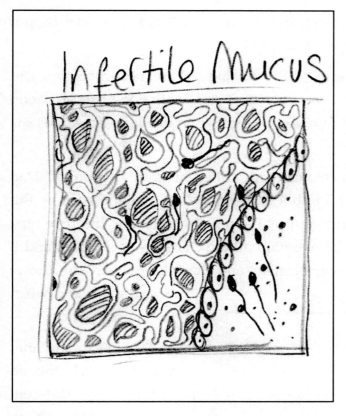

Figure 6: Infertile mucus

"And there you have a drawing of a very different sort of mucus. Can you see? It's full of blocks so that the sperm knock at it and get no further. That's how the cervix controls entrance to the uterus—the mucus changes with the phase in the cycle so that sperm either get through or are shut out.

"While this is going on, the corpus luteum continues to disintegrate until there's nothing left, so guess what happens next?"

"I suppose the hormone production stops?" suggested Michael, sneaking a look at the diagram. "And that somehow leads to menstruation?"

"Well done," Dad said. "Without the support of the hormones there's nothing to keep the uterus lining intact and so it falls away, causing the period to start."

"It's pretty ingenious," Michael mused.

"There's another diagram I want to show you," Dad said, shuffling through his papers until he found a graph. "There you are. If I tell you that hormones affect our moods, what d'you think happens to a woman with all that going on in her body?"

Michael looked at the graph and made a face.

"So if you find your sister a bit grumpy one day, give her space. When those hormones disappear at the end of the cycle and the period starts, a girl can feel quite out of sorts."

"Which means letting Josie get away with things!" Michael grumbled.

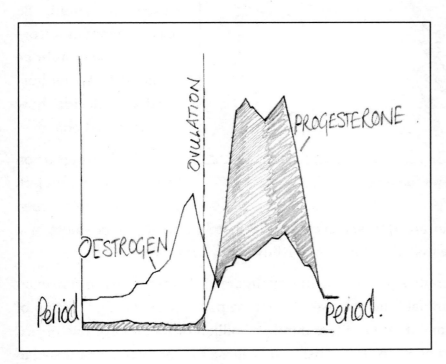

Figure 7: Female hormone cycle

"I'd be patient, if I were you. Hormonal swings in boys are even bigger than in girls, but they just don't happen on such a regular basis," Dad replied.

Michael had switched back to something else. "Dad," he said. "D'you remember you said that there are two big events in the cycle, menstruation and ovulation? Menstruation you—well, the woman—knows all about, but ovulation's going on inside so how does she know which bit of the cycle she's in?"

"That's a good question, because a woman's certainly different from an animal," Dad explained. "She doesn't go on heat, or give off public messages when ovulation comes. But she does have private signals. The most important is the mucus itself. The **infertile** type stays inside the cervix, but **fertile mucus**, the one with all the channels, is slippery and it drips down and out of the vagina. So the woman can see and chart it. That way she knows where she is and when to expect a period.

"There's another way in which human beings are different from animals. They don't just mate out of instinct—love always has to be free and making love is no different. It's something you can choose or avoid. So once a woman recognises that she might be fertile, she and her husband can decide how they want to behave.

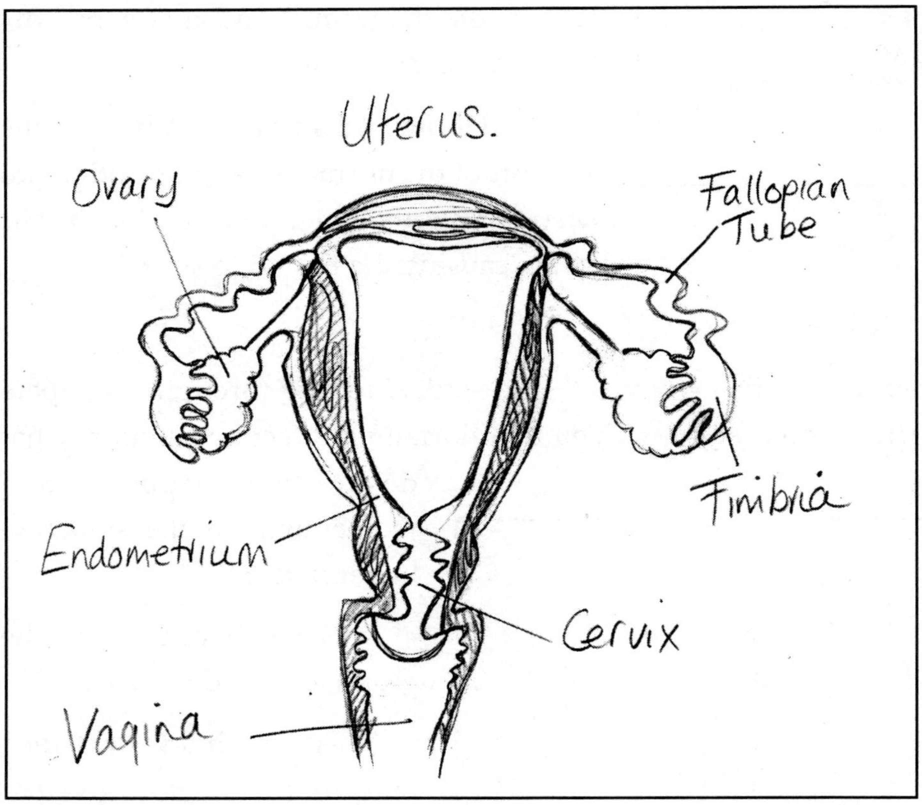

"Which brings us on to another subject, doesn't it? We've been talking about what happens in an ordinary cycle where there's no baby—and most of a woman's cycles are exactly like that. But the joy of fertility is having a baby."

He pulled out a labelled diagram of the woman's body (figure 8). "When a man and woman have sexual intercourse, the seminal fluid is released from the man's penis here, at the top of the woman's vagina. The sperm, all two to three hundred million of them, have one object, which is to reach the egg. But, most of the time, there's no egg, so the cervix remains closed to protect the female tract from unwanted foreign bodies.

Figure 8: Female reproductive organs

"But round the time of ovulation, and for a few days before, the cervix opens up a little and produces its fertile mucus. This doesn't just let the sperm swim through. It also helps to keep them alive."

"A bit like the seminal fluid?" asked Michael.

"That's right. In fact they work together. The mucus can keep the sperm alive inside the woman for up to 5 days, ready for the egg as soon as it appears."

"You mean you can make love one day, and fertilise the egg several days later?" Michael asked.

"Exactly so," said Dad. "It makes sense, if you think about it. After all, not many babies would be conceived if intercourse had to happen within hours after ovulation.

"The mucus has another function as well. It acts like a sieve and catches the many sperm which aren't up to scratch. What with the action of the mucus, and the obstacles of the female organs, it's only the very best sperm that get anywhere near the egg, a few hundred out of the millions that started out."

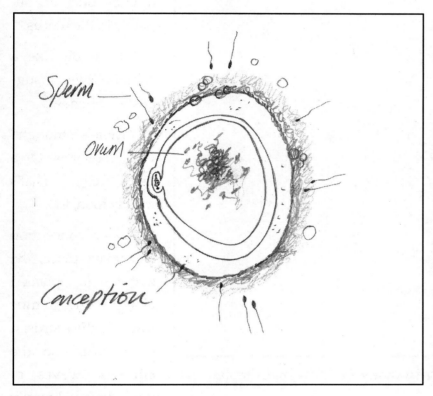

Figure 9: Conception

Michael was rifling through the other diagrams that Dad had brought down. "Is this the egg being besieged by the sperm?" he asked (figure 9).

"That's a good description," Dad smiled. "Oddly only one is allowed to penetrate. As soon as that happens, the outer layer of the egg seals itself off. It's a little detail I've always remembered. Nobody yet knows quite how it works."

"And that's what you call **conception**, when the sperm enters the egg?" Michael asked, examining the diagram.

"Yes, and no. It can take up to 24 hours for the sperm and the egg to fuse. And that's where you and I both started, when the egg and the sperm became a single cell. But what a cell! It already had in it everything that makes you you and me me—the colour of our eyes, the shape of our hands, even the way our hair grows."

"Dad, if the sperm's so small, and the egg's so big, doesn't the woman pass on more of her **genes** than the man?"

"I can assure you that you got as much DNA from me as you did from Mum. That first cell's actually called a **zygote**, which comes from the Greek for *being yoked together*. If you can imagine animals being yoked together, they have to work at the same pace, don't they? The reason the egg's so big is different. It's because it has all its own nourishment for the first 8 or 9 days of life. The baby only starts feeding from the mother once it implants in the uterus."

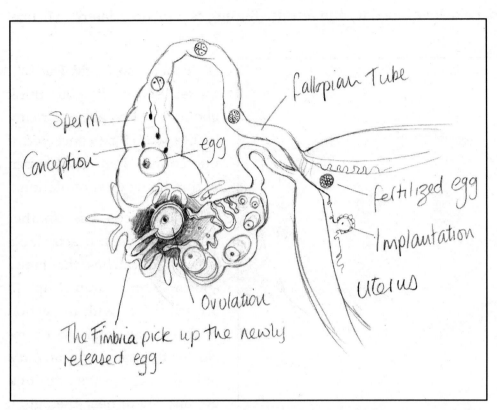

Figure 10: Embryo divides into many cells before implanting in the uterus

"It's a bit like a camel, then," suggested Michael.

"There's a thought! I suppose it is. Or a hen's egg. That's mostly food, too.

"Once conception has taken place, the zygote is pushed along the tube towards the uterus. It can't swim, so the **cilia**—or waves of tiny hairs—literally sweep it along. Then, nearly two days later, it starts to divide, into 2, 4, 8, then 16 cells etc, replicating its DNA as in normal cell division. By about day 6 it reaches the uterus. By then it's made up of hundreds of cells, but it's still the same size as it was at conception."

"Because it's had nothing new added? Like a hen's egg staying the same size even when the chick's ready to hatch?" Michael suggested.

"And I suspect it's getting pretty hungry by day 8 or 9 when it finally implants into the wall of the uterus," Dad laughed. "That's when it begins to feed and grow."

"Dad," Michael remarked, "our biology book says conception only happens when the egg implants."

"It does, does it? In that case it's following new terminology. Some scientists have their own reasons to think human life only starts at **implantation**."[3]

> **Twins**
>
>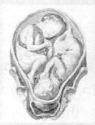
>
> **Non-identical or fraternal twins** come about when two mature eggs ovulate. They are released and fertilised by two separate sperm within the same 24-hour period. They each have their own **DNA**, and will grow up to be no more similar than any other brother or sister.
>
> Very occasionally, a single embryo divides into two within the first few days after conception. The single **zygote** thus develops into two babies, known as **identical twins**.
>
> The lives of both babies still start at conception, when the new 46-chromosome **zygote** is formed with its own **DNA**. The twins continue to share their **DNA** throughout their lives, which is why they look identical.
>
> But they are not identical in everything. Genes and the developmental environment in the uterus interact, so that the **DNA** in each twin is expressed differently. Once the twins are grown, a **DNA** test may not tell them apart, but a fingerprint test will.

"But you can see what they're getting at, can't you," Michael persisted, "if you can have one baby becoming twins several days after fertilisation? I mean, it's not very clear when the second life began."

"When neither life began was at implantation!" Dad exclaimed, gathering things up and moving back his chair. "The scientists give another reason, which is that many embryos get lost before implantation. They try to say it's OK if it happens artificially, for instance by taking drugs."

"That's ridiculous!" Michael exclaimed, getting up too. "As though the reason you die doesn't matter!"

Dad tidied up the table before looking at his watch. "We may have been quick with the washing-up," he remarked, "but we've made up for it in talking. Let's get outside while the sun's out."

* * * * * * * *

They decided to take a football up to the local park.

Dad remarked as they walked along, "All that biology gives you a very important pointer, which is that *sex is geared to giving life*. You can't get away from it, try as people might. You've seen the biology for yourself. In fact, it's hazardous even to think about sex without respecting its potential for giving life."

"Hazardous?" asked Michael.

"Yes, hazardous," Dad insisted. "You'll discover as you grow into manhood that the male sex drive can be very strong. You'll have every temptation to get physical with the girls round you, especially when they show interest in you. And you can expect to be teased for being restrained. However, sexuality is no game. It's not been given to you as a plaything, and if you use it like that it can bring a lot of distress to a lot of people, sometimes permanently."

"Like a child you might father by accident?"

"Yes, like a child. For a boy there's a special sorrow in fathering a child too young, because in most cases he loses it. You see, the relationship with the mother rarely survives the stress of pregnancy, which isn't that surprising, given that the relationship was immature and inappropriate in the first place."

"What about **contraception**? I mean, there are ways to stop pregnancy," Michael queried.

Dad shook his head. "One day I'll talk to you about contraception, but relying on it as a teenager is a mug's game." He paused, and then said, "Shall I tell you what it's like? It's like asking a girl to dress up in armour, armour which is known to be faulty, especially when youngsters use it. And then you fire at the girl at point-blank range with live ammunition. Who've you got to blame if the gun does its job and one of the bullets reaches its target?

"I'm not suggesting that a baby is a death—it's the opposite, it's a new life. But if you wait till you're married you can give life in an altogether more viable way.

"I'll tell you something else, which in a way is saying the same thing. And that is that *sex expresses permanent attachment*. It's so powerful that it actually changes you, too, even if you don't go as far as intercourse. Again, it's not to be played with before time, or it can cut across your friendships and mess up your ability to form a happy marriage later on."

Michael thought about this. He thought too about the discos he'd attended with his older cousins, and the way people treated each other at them. Dad seemed to read him because he said, "When I was a teenager, I used to think it unfair that our sex drive comes along so soon, and before a boy's in a position to think seriously about girls. Now I realise that this time of growing up plays its part. It helps your self-control."

"Just like training at football," Michael said.

"Exactly," said Dad. They were nearing the park now. "And you can't underestimate the need for training. You'll find that, as you grow bigger, there're bound to be occasions when the desire for sexual excitement attracts you just for its own sake. The danger even lurks when you're by yourself. It's easy to entertain sexual fantasies, or to excite your own physical responses. But if you **masturbate,** you're using your sexuality for your own pleasure and not in the way it's designed, which is *as a language of love for somebody else.* So my advice is, if you're tempted, get up, change your activity, think about something else."

Ahead of them, a couple of girls were chasing each other as they ran through the gate and into the park. Dad looked up at them as he added, "Your sex drive also gets you to take more interest in girls. You'll learn how to interact with them and make them happy, too. It's all part of the training for courtship and becoming a good husband. If you didn't learn when you were young, you'd be less good later on at coping with women emotionally and choosing the right wife." He bounced the ball across to Michael as they entered the park. "Come on. Enough of this talk. Let's see what sort of sportsman you are today!"

Points to remember

The woman's reproductive cell is called the egg (or **ovum**). Unlike sperm, which develop continuously throughout a man's life, the half million or so eggs which a woman has are already in her body when she is born.

Every cycle a hormone from the brain stimulates one of the two ovaries to select some eggs. These develop inside protective follicles.

One of them, the dominant follicle, grows faster than the rest until it fills and breaks open the ovary, releasing its mature egg into the abdomen. This is called ovulation.

The egg, by now about the size of a full-stop, is picked up by the fimbria and taken into the fallopian tube.

Conception, if it happens, occurs in the outer end of the tube. A single sperm out of the many released at intercourse is able to penetrate the egg. When the egg and the sperm fuse, at fertilisation, a new cell (or zygote) is formed, bearing the unique DNA which the child will have for life.

The embryo is pushed along the tube towards the uterus, which it reaches on about day 6. It implants in the endometrium on about day 8 or 9, where it takes in its first external nourishment. By then it has subdivided into many hundreds of cells.

Without fertilisation the egg dies within 12-24 hours. Some 10-13 days later, the endometrium is discarded, and comes away in drips of blood. This is called menstruation (or a period), and happens roughly monthly, with variations.

The cycle is controlled by hormones from the brain, which in turn trigger hormone production within the ovary.

Before ovulation, the growing follicle releases oestrogen. This makes the endometrium (or inner lining of the uterus) thicken into a 'nest', and makes the cervix start to produce fertile mucus, which appears round the time of ovulation.

After ovulation, the empty follicle (or corpus luteum) produces mostly progesterone. This maintains the 'nest' and makes it produce nutritious juices. The progesterone also acts on the cervical crypts, which switch to producing infertile mucus.

The cervical mucus plays a critical role in fertility. It lets sperm through the cervix or blocks its entrance, depending upon the time of the cycle. It can also prolong the life of the sperm for up to 3-5 days.

GLOSSARY

Cervix	Neck of the uterus, which controls access of sperm to the female's reproductive organs, especially by switching production between barrier and fertile mucus.
Cilia	Fine hair-like projections lining the tubes, which beat in waves hundreds of times per second to transport the egg.
Conception	The moment when a new human life comes into being, complete with its own unique DNA.
Contraception	Collective term for all artificial methods which intercept pregnancy before implantation of the egg.
Corpus luteum	Name given to the collapsed follicle. The corpus luteum produces the progesterone.
DNA	Deoxyribonucleic acid, which contains the genetic instructions for the development and functioning of all living organisms. As its name suggests, this acid belongs in each cell's nucleus.
Egg	Female reproductive cell, which, if fertilized by the male sperm, can develop into new human life. Girls are born with about half a million tiny eggs already stored in their ovaries. Also called by the Latin name **ovum** (**ova** in the plural).
Embryo	Baby in the first eight or so weeks of life after conception.
Endometrium	Inner lining of the uterus, which thickens each month to prepare an hospitable environment for the embryo to settle in.
Fallopian tube	Duct, of about 8 to 10 cm in length, which takes the egg from the ovary, and any sperm towards the egg. Also called an **oviduct**.
Fertilisation	Process by which sperm fuses with the egg to form the zygote. Fertilisation can take up to a day to complete.
Fimbria	Fingerlike extensions of the fallopian tubes, which catch the egg from the ovary.
Follicle	Cyst-like structure within which the maturing egg grows and which releases oestrogen and progesterone.
Hormone	Chemical messenger which carries instructions from one part of the body to another through the blood-stream. It can start a reaction, or block one.
Implantation	Attachment of embryo to the wall of the uterus.

Masturbate	To excite one's own sexual responses physically as a solo act. Masturbation also excites mental images, since the body, mind and spirit are closely connected.
Menstruation	Uterine bleeding, when the inner lining of the uterus is shed, so called because it usually follows a monthly pattern.
Oestrogen	Hormone which instructs the lining of the uterus to thicken, and controls the production of fertile cervical mucus. Also spelt **estrogen**.
Ovary	Organ which houses and matures the eggs.
Ovulation	Process by which the mature egg is released from the follicle and out of the ovary.
Pituitary gland	Small gland found at the base of the brain, which secretes the hormones which control the menstrual cycle.
Progesterone	Female sex hormone which maintains the endometrium and stimulates its glands to produce nutritious fluids. It also prompts the cervix to produce barrier mucus.
Uterus	A muscular organ, shaped like a pear, in which a baby develops and is nourished before birth.
Vagina	Elastic muscular canal which connects the cervix to the outside of the body.
Zygote	Name given to the single cell formed by the union of the egg and sperm and taken from the Greek word meaning 'to be joined or yoked'. The zygote lives as a single cell for about 40 hours.

Chapter 8

Changed for Life

"Dad?" Michael tried to sound casual, but actually he had been puzzling over his question for some time. "Dad, you remember we were talking about sex? Well, the subject came up yesterday and I didn't know the answer."

"The answer to what?" asked Dad. Michael had been at scout camp all week and they were both in holiday mood as they drove to catch up with the rest of the family.

"Well, when you can make love," Michael said.

Human fertility

A healthy man becomes fertile at puberty and, although his fertility diminishes as he gets older, he remains so for the rest of his life.

The changing pattern of human fertility therefore rests upon the woman, who is mostly infertile. She can only actually conceive in any one cycle during the 12-24 hours that a fresh egg is available to sperm. The fertile phase, when an act of intercourse can lead to pregnancy, is extended by fertile mucus, which can keep sperm alive in the female tract for 3-5 days.

Natural family planning methods may treat a wider span of days as potentially fertile, depending upon the woman's supply of mucus and other noticeable symptoms. It is easier to know when ovulation has happened than to predict its exact arrival.

A woman ceases to be fertile in middle-age, known as the **menopause**. Ovulation is also suppressed during pregnancy, and in the first months of breastfeeding. How long breastfeeding infertility lasts varies with the woman, her lifestyle and the way she feeds her baby.

His father was concentrating on the traffic. "You can make love whenever you want to, but most of the time it's at night or in the early morning, because that's when a husband and wife are relaxed and close and snuggled up in bed together." He stopped talking while he changed lanes. "Why do you ask?"

"Well, dogs only mate when the bitch is on heat," Michael replied, shrugging. "I wasn't sure if we might be like them."

"You've got a good question there," Dad replied. "Most animals, dogs included, only mate when the female is fertile, but men and women are different. They can make love at any time, when the woman is fertile, which is only for a few days every month, and when she's infertile, which is most of the time."

"Why's that?" Michael asked.

"You are full of questions today, aren't you? Still, they're important. The reason is that there's something very special about men and women. They don't just mate like animals. The act of intercourse really is designed for 'making love'. Making love uses the language of the whole person, spirit, mind and body, in a way which goes beyond words. It creates unity between husband and wife and helps to keep their love fresh."

"So how do you get yourself ready?" Michael asked. "D'you remember, like you were telling me last time? I mean, it would be a bit unfortunate if you thought you were going to make a baby and you went for a wee instead!"

Dad shook his head at this, and smiled. "I was explaining to you about the seminal fluid and the urine, wasn't I, and how only one of them can come through the **urethra** at any one time? I said it was a bit like railway points, switching from one track to the other." Dad paused for a moment as the lights went green and then, concentrating on the road again, said,

> **Neurochemicals**
>
>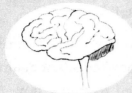
>
> Hundreds of chemicals travel about the brain's cells all of the time. Some carry messages. Others play critical roles in our thinking, desires and behaviour.
>
> **Neurochemicals** are released by the body in response to physical, mental or emotional stimuli. They are released automatically in response to these stimuli, regardless of whether the behaviour is morally good or bad.

"Most of the time your penis is floppy. It just hangs down on top of your testicles. But sometimes, like when you wake up in the night or first thing in the morning, it's stiffer and points upwards, doesn't it?"

"Yes," said Michael, starting to redden.

"Well, if you want to do a wee, you don't go when it's pointing upwards, do you? I mean, it's tricky till it flops down again."

"Which is just as well!" laughed Michael.

Dad grinned. "Agreed. Well, your penis becomes stiff because it **engorges** with blood. When the penis stiffens the 'points' turn, making ready for sexual intercourse."

"So when the penis is floppy, out comes the wee, and when it's stiff, it's the sperm?" suggested Michael.

"I suppose so," agreed his dad, "though the **seminal fluid** isn't all sperm. Most of it, as I told you before, is actually a sugary liquid which transports the sperm and keeps them alive."

Dad went on to explain, "You might wonder how the 'points' are controlled. Our spirits, minds and bodies are knitted so closely that you only have to think actively about sexual attraction for the body to go on alert and give you an **erection**. It takes a lot more sexual arousal than that for the fluid to be ejaculated. But you can equally switch those thoughts off, especially when you've learnt to be self-controlled. Can you remember which of your organs governs your sexuality?"

"Your brain?"

"That's right. And it's really important that your brain always remains in charge of your body."

"So, when you make love, what comes out is mainly syrupy water, with billions of tiny sperm inside?" asked Michael after a pause.

"Millions. The billions of sperm are how many you make each month, but in any one **ejaculate**—that means the amount which comes out at a time—there're more like two to three hundred million. What I don't think I told you before is that most of the sperm are concentrated in the liquid that comes out first. In fact, the concentration can be so great that it's possible for a woman to become pregnant without the penis entering the vagina."

"Do you know anyone that's happened to?" Michael enquired.

> ### Visual attraction
>
>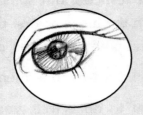
>
> Some of the most colourful displays in nature are designed for attracting a mate, such as a peacock's tail or flamingo's dance. Human beings are even more sophisticated, using all five senses to attract each other.
>
> What we see with our eyes can stimulate the **neurochemicals** in our brains, changing our moods and, on occasion, inclining us towards sexual arousal. A sunset over a still sea provides an atmosphere different from a rubbish dump on a dreary day. A smile from someone we love warms us, and a beautiful girl or a handsome boy attracts our attention and draws our gaze.

"What questions you ask!" Dad protested. "I do, as it happens. Though it's not the sort of thing one normally talks about."

"'How did you make love to your wife?'" laughed Michael. They fell silent again, but not for long. Dad wanted to know how the scout camp had gone, and Michael was regaling him with the latest jokes and how they'd been chased by a bull in a neighbouring field.

Then Michael remarked thoughtfully, "Dad, you always talk as though everybody who makes love is married. I know it's meant for marriage, but lots of people make love before that."

"Yes, I know. It's an easy thing to slide into. You can seem to be getting everything you want, the love and physical closeness, without having to commit to the long term. And sometimes there are even practical advantages to moving in with each other, such as sharing the bills."

"Like Annie and John?" suggested Michael. "Annie was round at our house the other day and she told Mum that renting a one-bedroom flat was the only way they could afford to live in town."

"Starting out in life can be difficult," Dad said, "and it can be tempting to shortcut the waiting in a relationship. You'll find that, even when you love a girl, you'll hang back from being tied to her before you're able to get yourself established. Marriage is a big step, and young people need to have money and stability before taking it. But waiting takes self-control and sex is very tempting."

"If you live together you might have a baby," Michael pointed out.

Dad smiled ruefully. "You're right," he replied.

"What about all those ways to stop a pregnancy, like the condom and the Pill?" Michael asked.

"And other things too," Dad replied. "But **contraception** is a big subject in its own right. I'll tell you now that it has a high failure rate, especially among young people. If a girl messes with it when she's young, she may damage her ability to conceive in the future."

"At least that doesn't affect a boy," said Michael.

> ### The brain
>
>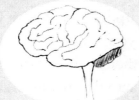
>
> **Neurons** are the primary cells of the brain. Electric impulses connect them, which is how the brain works. Each **neuron** has several short projections, called **dendrites**, which receive transmissions, and one long projection, called an **axon**, to send them.
>
> The **neurons** are not 'wired' to each other but are linked by **neurochemicals** to connectors, called **synapses**, through which messages are passed. The synapses are created, strengthened or killed off according to the use made of them (or lack of it). The **neurons** form over 100 trillion connections with each other—more than all the Internet connections in the world.
>
> It is the impermanent nature of the **synapses** which allows the brain to be continually moulded throughout life according to how we behave, think and feel.

"It does if it's your future wife," Dad reminded him. "And boys can catch **sexual diseases** as much as girls."

They had caught up with a large lorry which was going at just the wrong speed to overtake. At last Dad was able to accelerate past it.

Michael had taken out his mobile and was fiddling with it. Without looking up he remarked, "Anyway, when I marry I'll want to have some way of saying 'I love you' which I haven't already used up."

"You know, your age group has got such a chance to learn from our mistakes," Dad replied.

"I told you just now that our spirits, minds and bodies act in such unison that you only have to think actively about sex for your body to react. The way this happens is through chemicals in our brain (**neurochemicals**), which link our thoughts and emotions to our physical reactions. Brain scan technology is teaching us more about this all the time. It seems that sexual activity gives rise to potent chemicals which have a huge influence on our behaviour. We're designed not only to fall in love with another person emotionally, but to become physically attached to that person through chemical attraction, almost like an addiction."

"So that's what you meant when you said that love-making creates unity?" Michael replied, impressed.

"It's even more than that. We don't just become chemically bonded. Making love creates new electrical circuits in our brains which change the way we think. Didn't you tell me that you learnt at school about brains and how they work?"

"Yeah, we did," Michael agreed. "It was when we were learning about sport and they told us that exercise makes new connectors in the brain. They showed us how we build our brains at the same time as our muscles."

"OK," Dad responded. "Well, sex builds your brain even more than sport. You'll remember that the connections between the **neurons** aren't fixed like a wire, but that the electricity jumps from one neuron to the next? They use connection points called **synapses,** don't they?

> ### Dopamine
>
>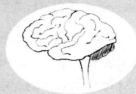
>
> This neurochemical has many functions, including that of giving a feeling of intense energy and exhilaration when we pitch into demanding activities or take risks. The number of **dopamine** receptors in the reward centre of the brain declines in adolescence so that young people need high levels of excitement. This encourages them to be bold. The **dopamine** reward creates habits of behaviour, which can be either good or bad.
>
> **Dopamine's** release can also be artificially triggered by stimulants such as alcohol, nicotine and recreational drugs. When overstimulated, the brain learns to become resistant, needing more for the same effect and so creating addiction.
>
> Sex is one of the strongest generators of the **dopamine** reward. A couple in love are excited and confident, and prepared to take risks so that they can be together. The sexual act then seals their love and keeps them loyal to each other. Sex indulged in lightly can trap people into wanting it for its own sake, in increasing quantities.

And where the energy runs, pathways are either created or shut off, and that depends upon how you're thinking and the way you act. Now, when you bond with another person sexually, you create a lot of activity in the brain—so much activity, in fact, that it has a really big impact and your brain is never the same again.

"Being bonded to your husband or wife is wonderful. Then your sexual makeup is matched by your actions—all those little things you do for each other throughout the day, the responsibilities you take on, the goods you share, the way you rely on each other, especially when you have children. That's real love, even when you don't always feel it as 'being in love', because feelings mature over time."

"That sounds a bit dull," Michael said.

Dad chuckled. "It's also a lot more peaceful, and it gives you a deep sense of well being."

"Which you don't have till you marry?"

"Especially when you're young," Dad replied. "You may not even have chosen your partner with much care. So along comes an obstacle and the relationship breaks up. But the chemical bond is still there, unsatisfied. You are left with a taste for sex, which you will look for again, but this time you'll be tempted to jump in even more quickly.

"The sad thing is that you can end up wanting sex for its own sake, and lose the ability to bond in a long-term relationship. It's a bit like a plaster. It's designed to stick once. Try it again, and it might stick, but third and fourth time round it sticks less and less. But people aren't plasters, and when they're torn off their self-esteem takes a knock, sometimes a big one. You'll find that teenagers who get into sex start losing grades at school, and they're much more likely to drink and smoke and get into the drug scene."

> ## Oxytocin
>
>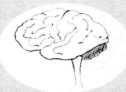
>
> This **neurochemical** is primarily active in a woman, and bonds her to her sexual partners, and to her babies.
>
> Any meaningful sexual contact can begin the release of **oxytocin**, even a hug. Its effects are so powerful that it can cloud a woman's judgement and make her desire and trust a man to whom she is unsuited. When sexual relations are reserved for her husband, **oxytocin** helps a woman to overlook his faults and enjoy a stable, well-bonded marriage. The further release of **oxytocin** at childbirth and in breastfeeding cements family relationships.
>
> The breakup of a relationship is altogether more traumatic when full sexual bonding has taken place. It weakens the ability to bond satisfactorily with another man in the future.

"Dad," Michael asked, "if sex is such a big deal, why isn't everybody warning us off it?"

Dad sighed. "Sex is big business. Put a gorgeous girl into an advertisement, and it sells a car. Include a sex scene in a film, and the audience goes up. And that's to say nothing of the money that companies earn from pornography and contraceptives."

He paused for a moment, concentrating again on driving, then said in a brighter tone, "Now it's my turn to ask the questions. Did you know that men who are married to the mothers of their children can expect to earn 20-30% more than men living in a similar situation but without being married?"[4]

"Really? Why's that?" exclaimed Michael.

"I'll leave you to work that one out. But it centres on security and long term commitment—and generally knowing that your family is fully yours for life." They drew up at some road works and Dad, pulling on the brake, relaxed back into his seat. "And I've got another one for you. This one's more difficult. Have you ever thought about marriage and what it really is?"

"It's a solemn promise to stay with someone till one of you dies," Michael returned.

"You're right: it's a very solemn public promise, and because it's witnessed by the state, and often by a religious authority, it makes it easier to keep. But apart from promising to stay with someone for life, what do you also promise?"

Michael looked a bit puzzled.

Dad helped him out. "Well, you promise to be faithful, don't you? Marriage is exclusive—just for two people—and it's forever. Now, let's get back to the act of intercourse. The language of the body in having intercourse is saying exactly that: 'I love you so much that I want to share everything I have with you, even to my seed.' The seed means either the man's sperm or the woman's egg. There's nothing more solemn or more exclusive and long lasting than sharing in giving life to a new human being. So whether or not a child is

> ### Vasopressin
>
>
>
> This neurochemical is the male equivalent of oxytocin. It has many functions, but is specifically responsible for bonding a man to his sexual partner and to his children. Like oxytocin, it is designed to support a stable long-term relationship within marriage.
>
> Men, like women, are susceptible to losing their judgment because of premature physical bonding. Their brains are flooded with **vasopressin** each time they have sexual intercourse, producing a partial bond with each woman. Bonds thus made and lightly broken make it more difficult for a man to commit himself in marriage.
>
> Men who have many partners depend for their sexual satisfaction upon the dopamine rush. They also mould the neurological circuits of their brains into accepting multiple partners as normal.

actually conceived, the giving of the seed by the man to the woman is a sign of lasting and exclusive love. It belongs to marriage.

"So what do you think happens when you make love without that commitment?"

Michael had to think hard. "I guess it's dishonest," he said slowly, "because your body is doing one thing while your spirit is saying 'I don't really mean this'."

"That's very well put," Dad responded. "I may say that, at the time, the actions of the body can make you feel so good and so close that you can believe you are giving everything."

"That means, surely," Michael said, "that you're cheating yourself as well as the other person."

"And I am warning you, it's very easy to do."

* * * * * * * * *

A few weeks later Dad was driving Michael back to scouts. It was the first meeting of term and he was normally excited before seeing everyone again.

"You're unusually quiet today," Dad remarked. "What's up?"

"Nothing," Michael replied. Then he said, "Dad, you remember last time we had that conversation about sex? Well, what happens if boys are attracted to each other? I mean, do they have to wait till they're grown-up? After all, since you can't have a baby, surely it doesn't matter when you start having sex."

Dad was taken aback by this unexpected question, but answered calmly, "What makes you ask?"

"I saw Ben last week, and he told me that, at the scout camp, he was sharing a tent with Kevin, you know, the boy who moved in down our road? He's a really quiet boy, keeps himself to himself most of the time, but as they were in the tent together they started talking. And Kevin told Ben that he's in love with Johnnie Campbell. He keeps fantasising about him,

> ## Prefrontal Cortex
>
>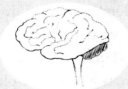
>
> Our capacity for cognitive thought, when we make judgments, sort out priorities and control our impulses, is seated in the **prefrontal cortex**, just behind the forehead. This most sophisticated organ ultimately connects with many other areas of the brain. It only reaches maturity at the age of 23-25.
>
> Other advanced brain organs also mature in the early 20's. These include the **amygdala** (emotion centre) and the **hippocampus** (memory centre).
>
> By the time the **prefrontal cortex** is fully developed, most young people have already completed their studies and decided upon their adult occupations. Many have experienced their first loves, and may even have married or set up home together.
>
> Young people have always turned to older ones for advice and we can now see that it is not only because adults have more experience: they are physically capable of better joined-up thinking.
>
> Car insurance companies know this. They regularly charge higher premiums to the under 25's.

and blushes whenever Johnnie speaks to him. He feels really embarrassed and thinks we must all notice, and that makes him even shy of us too. He thinks he must be homosexual and doesn't know what to do about it. Which just made me wonder."

"So, there are two questions there, aren't there?" Dad pointed out. "The first is about active sexual behaviour with somebody of the same sex. And my answer to that is the same as if you were attracted to somebody of the opposite sex. Sex binds you to another person, but it also carries with it the symbolism of new life, doesn't it?"

"You mean that business about the man giving his seed to the woman? But this would be giving it to another man," Michael mused.

"So if the symbol of life isn't appropriate, which it isn't, the sexual act isn't appropriate either. Look at it another way. If I had x-ray eyes, I would be able to understand the function of your liver, and your heart, and your kidneys, because they're all self-contained inside your body. I could see them cleaning and pumping your blood around. But what explanation would I have of the testes, or the prostate?"

Michael puzzled over this. "Well, they produce the sperm and the seminal fluid."

"But what's the point of them? What purpose do they have? You can't explain sperm by just looking at the man's body, can you? They only make sense in the context of a woman. And the same with all the woman's reproductive organs, the ovaries, and the tubes and the uterus. It's only in the sexual act, when the man and woman join together, that you can make sense of either body by itself."

Michael thought about this hard. "So it's like the two bodies become one?" he asked, clearly impressed.

> ## Pornography
>
>
>
> **Pornography** is the explicit portrayal of images which cause sexual arousal. It can be through any medium: the internet, magazines, films, animations, or on any electronic device. It is dangerous and very addictive, setting off patterns in the brain different from those experienced when looking at the beloved.
>
> Increasingly extreme images are needed to maintain the same level of excitement. The user's own sexual life can appear insipid by comparison so that the user loses the enjoyment of ordinary sexual relations. Even previous **pornographic use** interferes with the ability to bond well with one person.
>
> **Pornography** imprints itself on the brain, leaving lasting impressions in the imagination. These can pop up again at any time; images viewed on screen are especially dangerous.
>
> **Pornography** also exploits the models, portraying them in poses which attack their dignity as full human beings, composed of mind, body and spirit. Often the models are vulnerable people taken advantage of because of their youth or their financial and social need. **Pornography** is a big and growing business making billions of dollars every year.
>
> ### If a sexual image comes up, don't wait: turn it off!

"Yes, it is," Dad replied, "and seeing the body like that will help you understand something else I told you, too, which is why **masturbation**, or engaging in sexual activity on your own, is also out of place. You see, the seed of your body is precious, and it has to be treated with respect." Dad could see Michael's expression tautening at this. He went on lightly, "It's common at your age not yet to be in full control of your body. It's just something to be aware of and to work on.

"It's important to bear in mind that none of us has an automatic right to something just because we're attracted to it, in sex or anything else. And those with same-sex attraction aren't the only people asked to control their desires. Everybody does, whether married or single.

"And you might be surprised at quite how many people do live without engaging in sex. Some choose to do so because they have a specific religious vocation, or because they're happier being independent. But there are many others who lead a continent life without choosing it. Some have never found the right person to marry. Others are widowed, or divorced—and many of these had separation forced on them. Whatever the case, luckily you don't have to be actively sexual to enjoy firm bonds with other people.

"But there's a second part to your question, which is what you do about Kevin. It sounds as though he's feeling a bit lost and is looking for acceptance in the wrong place. There are lots of reasons why he might be feeling as he does, but, at the nub of them, there's likely to be something which tells him that he doesn't quite measure up as a boy. It's not unusual to muddle up the desire to be loved with sexual feelings. Most young people settle down if they're encouraged in the right way.[5]

"So the best way for you to help Kevin is to make sure he feels a boy among boys. Why don't you invite him home? And get Ben and the others to include him in your group?"

Michael raised his eyebrows at this. Typical Dad to suggest that he should adopt some boy he didn't even like. But then perhaps he didn't really know him. And he did live up the road.

Kevin was walking along the pavement as they arrived. Michael got out of the car, and then with a set face signalled to Dad to wait while he hurried after him. "Hi, Kevin," he called, catching up. "My dad's asking how you're getting home tonight? Would you like a lift?"

Points to remember

The penis conveys both urine and seminal fluid but never at the same time.

Seminal fluid is made up of sperm and the sugary liquid which conveys, nourishes and protects them. The greatest concentration of sperm is in the liquid which comes out first.

After puberty, a healthy man remains fertile for the rest of his life. A woman is mostly infertile: her fertility is restricted to 3-5 days in any one menstrual cycle and ceases in middle age.

Unlike the vast majority of animals, men and women can have intercourse at any time in the female cycle, regardless of whether or not the woman is fertile.

Every full sexual act retains the symbolism of giving life since the man releases the seed of his body. Sex is only truly satisfying when the language of the body is in harmony with that of the mind and spirit.

The brain governs all our sexual functions and is the link between the physical and the spiritual. We can exercise control over our sexual responses.

Neurochemicals prompt sexual attraction and are released at intercourse in such numbers that they impact upon the connections in our brains, changing the way we think and behave. This is designed to bond us permanently to our sexual partners.

A man can only fully understand his masculinity in the context of a woman, and a woman her femininity in the context of a man. In the sexual act, the man's and the woman's bodies function organically as though they were a single body.

It is not uncommon for adolescents to be ambivalent about their sexual identity and to have same-sex crushes. The best way to help them is to affirm them through the warmth of friendship, from both girls and boys but especially from people of their own sex.

It is not necessary to engage in active sexual behaviour in order to enjoy close bonds with other people.

Pornographic images leave lasting impressions in the brain and are highly addictive, especially when viewed on screen, for instance over the internet. They interfere with the ability to enjoy ordinary sexual relations and to bond well with a spouse.

GLOSSARY

Amygdala — Almond-shaped groups of nuclei which perform a primary role in processing and storing emotional reactions.

Axon — Long projection attached to the neuron for sending transmissions.

Brain scan — Images the brain's activity by measuring its points of energy release and blood flow. Developed in the 1990s, it is allowing scientists to study the brain's reward pathways.

Cervical mucus — Produced by the woman's cervix round the time of ovulation to help the sperm reach the egg. It also nourishes them and prolongs their life.

Contraception — Drugs and devices used to avoid pregnancy.

Dentrites — Short projections attached to the neuron for receiving transmissions.

Dopamine — Chemical released in the brain, which rewards exciting or demanding behaviour with feelings of intense satisfaction and pleasure. It plays an important role in many aspects of human behaviour, including the sexual. Dopamine release can be over-stimulated with the use of drugs, alcohol and casual sex, leading to addictions.

Ejaculate — Ejection of seminal fluid from the penis. The average ejaculate has a volume of 3–5 mls, containing some 200 to 300 million sperm.

Engorges — Fills up with blood. The penis contains lots of tiny blood vessels which rapidly fill during an erection, causing the penis to stiffen.

Erection — Stiffening of the penis through a surge of blood, which fills up the many extra blood vessels which the penis contains.

Hippocampus — Major component of the brain's limbic system, which consolidates short- and long-term memory and spatial navigation.

Masturbation — Exciting one's own physical sexual responses as a solo act. Masturbation also involves sexual fantasies since the body, mind and spirit are closely interconnected.

Menopause — Phase of life, usually in her 40's or 50's, when a woman ceases to ovulate.

Neurochemical — Chemicals which bathe the brain cells and carry messages across the synapses from one cell to the next. Neurons form over 100 trillion connections with each other, more than all the Internet connections in the world.

Neurons — Primary cells of the brain through which electricity flows to make the brain work. By the end of adolescence, a person's brain will contain more than 10 billion neurons.

Ovulation	Release of the egg from the ovary into the fallopian tube, where it becomes available for fertilisation by the sperm.
Oxytocin	Neurochemical, especially active in women, which stimulates feelings of bonding towards a man and towards her child. Oxytocin is released in large quantities during the sexual act, at childbirth and in breast-feeding.
Pornography	Explicit portrayal of images which cause sexual arousal.
Prefrontal cortex	Area of the brain responsible for executive functions, such as defining goals, setting priorities, assessing consequences and controlling impulses. It only matures at age 23-25.
Seminal fluid	Fluid ejaculated by the penis and made up of sperm and the water and various nutrients which carry the sperm. The seminal fluid contains a concentration of fructose, a very sweet sugar which feeds the sperm and gives them energy.
Sexual diseases	Diseases transmitted through sexual contact with a carrier. They are usually referred to as STDs (sexually transmitted diseases) or STIs (sexually transmitted infections) and are particularly catching among young people, whose defence mechanisms are not yet fully mature.
Synapses	Connectors which help carry electrical messages from one neuron to the next. The brain is not hardwired: the gap between the neurons and the synapses is bridged by neurochemicals. Synapses form and disintegrate according to use, making the brain continually adaptable, but especially so during adolescence.
Urethra	Duct which takes either seminal fluid or urine through the penis to the outside of the body.
Vasopressin	Neurochemical which functions in a man much as oxytocin does in a woman, bonding him to his mate and to his children.

Chapter 9

"Dad, Is Sex Dangerous?"

"What did they want you up at school for?" Michael asked as his dad came up to say goodnight.

"The Head wanted all the parents of your year to hear …"

"Come on, Dad. It was somebody from the hospital, wasn't it?"

"Yes, it was a man called Dr Peterson, and I think he gave most of us quite a shock."

> ### Sexually Transmitted Diseases (STDs)
>
>
>
> are infections passed on through sexual contact. You can't pick them up by sharing a glass or shaking somebody's hand. But you don't have to have intercourse to pick up an **STD**. Other intimate sexual activities are enough.
>
> **STDs** can be caught in a first encounter. And if you acquire one, you are more prone to catching others as well.
>
> **STDs** are also called **STIs (Sexually Transmitted Infections)** to emphasise that they can be present even when there are no obvious symptoms.

"Why's that, Dad?" asked Michael, getting interested and realising that here was a chance to delay lights-out.

His dad sat on the bed. "He was talking about **STDs—sexually transmitted diseases**."

"You mean **AIDS**?" asked Michael.

"I wish it was only AIDS. AIDS is the worst infection, but it's very rare among school children in this country. No, there are many others."

"Dad," said Michael. "Why did the Head have to call all the parents in? I mean, surely you learnt about all these things when you were my age? Why do you have to learn about them again now?"

"When I was your age," Dad replied, "school children knew next to nothing about STDs. We used to hear about AIDS and the wise guys might have known the words **gonorrhoea** and **syphilis**, but nobody really knew much about them, and we certainly weren't taught about them at school."

"Dad, is sex dangerous?"

"Funny you should say that," Dad replied. "That's exactly what Dr Peterson was telling us. Sex has become really dangerous." He shook his head sadly. "But it doesn't have to be that way. In fact, when it's treated properly, it isn't dangerous at all. It's completely natural—something our bodies are made for. It also gives great happiness. Let me explain.

"I have already told you enough about sex for you to know that it has two purposes. The first is to conceive children—without sex the human race would die out."

"And we'd be the last children left," put in Michael.

"Exactly! What a grim world that would be. The second purpose is to form a loving bond between the man and woman, so that they want to stay with each other for the rest of their lives. Saying 'I love you', not just in words but with your whole body, is as close as you can get. It strengthens the relationship between husband and wife and keeps it fresh.

"And, if you obey the body's signals, there's no reason why there should be any problems. For instance, black men can marry white women, Indians can marry Chinese. But if you marry someone from your own close family, the children lack the necessary genetic mix and could suffer, so you'll find that in every society sex between close family members is illegal."

"As if I would want to marry Josie!" exclaimed Michael, pulling a face.

Dad smiled broadly. "And I don't suppose she's lining up to marry you either. We don't want to marry our brothers and sisters, and as a result we can have a really good friendship with them all our lives."

Michael had a thought. "So, nature tells you that brothers and sisters are not meant to marry because their children would be less healthy?"

"Yes," Dad agreed.

"And nature also says if you have sex with different people you'll start getting diseases?"

"You run the risk," Dad said. Then he went on, "I had never appreciated before quite how unusual the reproductive process is. The body's ordinary defence mechanisms have to be suspended in a remarkable way for this one activity. Dr Peterson pointed out that people usually don't even share a glass for fear of germs. However, when a man and a woman make love they share much more than a glass.

"But it's not only germs which are passed between them. Since the man's sperm is foreign to the woman, so also is the baby. If there's ever going to be a successful pregnancy, the woman's body has to learn to accept it."

"What do you mean by that?" Michael asked.

"The baby gets two different sets of DNA, doesn't it, one from the mother and one from the father? So the mother's immune system has to accept that part of the baby's DNA that comes from the father. The cervical mucus and seminal fluid both play their part, but so do the sperm themselves.

"Dr Peterson told us something quite extraordinary. He said that all the myriads of sperm which the husband passes to his wife in sterile acts of intercourse help prepare the woman's immune system. Even during a fertile act, the winning sperm relies on its mates to achieve conception. As a result, men with a low sperm count find it more difficult to father a child."

Michael thought about this for a moment. "What about a honeymoon baby?" he questioned. "The woman's body wouldn't have prepared for that, would it?"

"Dr Peterson didn't say that healthy pregnancies *can't* happen first time round, but sometimes it takes longer for them to happen at all—rejection of the man's sperm is a cause of infertility. First pregnancies are also associated with a higher risk of **pre-eclampsia**, which can be very dangerous for both mother and child. It's a risk which appears to affect regular condom users, too. In either case, the wife doesn't have the same chance to 'get to know' her husband's sperm."

Sexually Transmitted Diseases (STDs)

are commonly caused either by **bacteria** or by a **virus**. The two are very different.

Bacteria are simple single-celled organisms which can live and multiply on any surface, living or otherwise. Bacteria come in many kinds, many of them performing useful functions. Only some are harmful. These bring disease by releasing toxins and rapidly increasing in number through cell division.

Viruses are many times smaller than a cell and are always harmful. They can be thought of as loose genetic information searching for a host cell to penetrate. Once in, the virus takes over the cell's mechanisms or integrates itself with the cell's own DNA. When enough 'baby' viruses have been formed, the cell bursts, releasing new viral particles.

"Dad," said Michael thoughtfully. "D'you realise what that means? It means that making love is always connected with having children, even when there isn't a chance of a baby."

His father caught his eye, clearly impressed. Then he resumed seriously: "Unfortunately, you can't have it both ways. If the woman's defences are down for the sake of a possible conception, they're also lowered against infection during the fertile period and during a subsequent pregnancy. Normally the vagina is very acidic and kills off germs—and sperm too. But, around the fertile time, the mucus transforms the vagina so it's easier for both to thrive.

"Dr Peterson then told us about STDs. He went out of his way to tell us that lots of his patients have caught an infection during their first sexual contact. There are now so many infections circulating."

"But surely young people don't catch them much, do they?" Michael asked. "I mean, most of us are fit and we'd throw off an infection fast enough."

"I'm afraid you're wrong there," Dad replied. "Children are *more* at risk of catching infections than adults, including sexual ones. The reason is that it's only when you grow into adulthood that your immune system is in full working order. The same is true of the reproductive system. That only comes to maturity at the end of adolescence, particularly in girls."

Michael was temporarily silenced by this. Eventually he said: "So if you get one of these infections, what happens?"

"The first thing is to know you've got one. As I said, there are lots around now. Dr Peterson said the first thing you'd notice would be discomfort or pain passing urine. Then there could be pain or irritation round the genital area, swellings, ulcers, unusual discharges, that sort of thing."

Dad noticed that Michael had gone quiet. "What's up, Mikey?" he asked. "Have you noticed something like that?"

Treatment of Bacterial Infections

STDs caused by bacteria can be cured by antibiotics if they are caught soon enough. This includes infections such as chlamydia, gonorrhoea and syphilis.

But you can't treat a disease you don't know you have. Both bacterial and viral STDs tend to hide their early symptoms so that you only know you've caught one when the damage is done. STDs are a major source of later infertility.

The only way of knowing if you have an STD is to be tested—or to wait until the symptoms are inescapable.

"Not really. But sometimes it does feel a bit itchy. Do you think I've got an STD?" Michael asked, his eyes growing wide.

"Only if you've had sexual contact with another person," Dad replied.

"But I could have given it to myself?" Michael said, looking dismayed.

"No way!" Dad reassured him lightly. "STDs can only occur during sexual activity with someone who is already infected. You're much more likely to have got a bit of irritation under your **foreskin**, but if you roll it back when you're in the bath or shower, it'll clear up. As for girls, they are more at risk of urinary infections before their menstrual cycles are fully established.

"Dr Peterson said that it's not unusual for patients to come to the clinic worried about an STD even though they've never had sex. The real problem, he said, is that too many people don't turn up who should be there. As with everything else, the sooner they're treated, the better."

"So these STDs can be cured?"

"It depends on the disease. They can all be treated, but some can't at present be cured. It seems there are various agents that cause STDs, the chief two being **bacteria** and **viruses**. The bacterial ones can be cured with antibiotics if they're caught soon enough, but they haven't yet found a way of curing some of the viral ones. The symptoms can be treated, and sometimes disappear of their own accord, but with some infections the virus stays in the system for life, and can be passed on to other people."

"You mean you could be saying, 'Please marry me, and I'll give you a disease.' Wow. That's cool, isn't it?" said Michael sarcastically.

"You can laugh, but it's no laughing matter for those caught out. Think of their dilemma. They know they will almost certainly infect a future husband or wife, so it's a real test of character as to how they act. It seems there are many young people in this position.

"They're angry, too, about having become infected. There's nothing they can do, and as often as not they don't see why anyone should know. So they ignore the problem, and pass the infection on to their next partner. You can imagine what that does to a relationship. Dr

Peterson told us that what saddened him most about STDs is how they change people's characters. It's not only in their bodies that some people are marked for life."

"If the symptoms disappear after a bit, it can't be that bad," said Michael, looking at things more optimistically.

"Yes and no," Dad replied. "The real problems start when there aren't any symptoms, because then the infection isn't treated. And if the infection remains, it can have serious consequences. One of these is infertility. It may not seem very important to young people, but when they're older it will. Imagine how a couple will feel if they can't have children because **chlamydia,** from a previous relationship, has blocked the wife's fallopian tubes.

"Chlamydia is the most common STD, and the majority of infected people don't know they have it. Apparently this wasn't always the case. Some years ago, according to Dr Peterson, infected males usually had enough symptoms to make them turn up at a **GUM clinic**. The patient would then be given contact slips to pass on to their female partners, inviting them to attend as well. But in the last ten years, men are also presenting with few to no symptoms, so it's harder to pin down.

"The reason is that bacteria, like all other organisms, are programmed to survive. You've probably heard people saying that bacterial infections are getting harder to treat, because they're becoming immune to antibiotics."

"You mean," said Michael, "if lots of antibiotics are used to cure STDs, they'll also become less effective against some completely different illness?"

"Alexander Fleming—you've probably learnt at school that he discovered penicillin?—well, he warned that antibiotics would become less effective the more they're used, and we're beginning to see it now. There are also some viral infections—which are quite different from the bacterial ones—which remain in the body even after treatment, things like **herpes** and **genital warts**."

"How horrid!" exclaimed Michael. "But if there aren't any symptoms, how do you know you have an infection?"

"You don't, without being tested. That's why Dr Peterson's message was simple, 'If you've been exposed to the risk of infection, get yourself to an STD clinic fast'."

"And what counts as risk?" Michael asked.

"It means—nowadays—any form of sexual contact with another person," Dad replied.

"What about **condoms**?" Michael persisted. "I thought they were meant to protect people?"

"It's true that condoms help against some infections, but they do little to nothing against others.[6] They were originally used to avoid pregnancy, but everybody now recognises how often they fail among teenagers—and conception can only happen periodically, whereas disease can pass at any time. The problem with condoms is that they can give people a false sense of security, leading people to disregard the dangers of sex. Then they're more promiscuous and catch more infections.[7] There is, of course, one fail-safe way not to get one …"

"… and that's not to have sex, right?" put in Michael. "Well, I don't want to anyway."

"Not right now, I hope," answered his father. "But you'll find that sex isn't like drink or drugs or smoking. Those are all outside of you and they need never get a hold of you, if you don't let them. The sexual drive is on the inside. It's natural to everyone, it's beautiful, and it can be very powerful indeed. It has to be, or men and women would never surrender their independence to make new families. And where would you be if your mother hadn't won me?"

Michael's dad looked at his watch. "And you've caught me, too, haven't you? Look at the time. Lights out and get to sleep quickly, or Mum will be on to me." He gave his son a kiss and made for the door.

* * * * * * * * *

The following day, Dad was having a late tea when Josie wandered in. She took an apple and joined him at the table.

"Hi, Dad," she said, taking a bite.

"Fancy seeing you!" Dad exclaimed. "I thought you were out with the others."

"Had too much homework," she replied. "Anyway, I'm going to Beth's tonight." She paused for a moment. Mum had done her best to pass on to Josie what Dad had learnt from Dr Peterson, and it was continuing to disturb her. She plucked up her courage and said so.

"Nothing like that need ever happen to you," he reassured her. "Just knowing the dangers is part of making sure it never does."

"Well, it's enough to put you off sex forever!" announced Josie, hunching on her stool.

Dad let the comment pass. He finished his tea purposefully and put down his mug; he was going to have to deal with this one carefully. "You can't ignore sex, you know, because it's part of who you are as a girl," he remarked. "Even if you never go out with any boy, and never marry — which could happen — you're still a sexual being, and the way you lead your life will be marked by that."

"How come?" Josie asked, taking another bite of the apple.

Dad leant over and gave her a sudden pinch.

"Ouch!" Josie exclaimed, drawing back. "Dad, what are you pinching me for?"

He chuckled. "I was just proving a point. If I hurt any part of you, it's you I hurt, isn't it? So you can't think of yourself as being separate from your body. That means that your body speaks about who you are as a person."

"Saying I'm a girl inside too?" asked Josie.

"Have you ever thought how instantly recognisable men and women are? You can usually tell them apart even at a distance, or over a telephone. That says that there's more to us as sexual beings than just needing two people to bring up children."

"Like what?" asked Josie.

"Well, forget preconceived ideas and look at the way men are and the way women are. You can learn a lot about how to make the best of your life as a woman from observing human nature honestly."

"Such as avoiding casual sex so that you don't get STDs?"

Dad frowned at that. "Some people get an STD first time round, Dr Peterson emphasised, and some spouses catch them innocently from a husband or wife who has been unfaithful. It's also possible to catch something like HIV from a dirty needle, or a blood transfusion, or even from a parent. But broadly speaking, yes, STDs warn one off promiscuous behaviour.

"However, I was actually thinking of something more general than that. The way men's and women's bodies interlock physically, creating one reproductive system, gives you a good clue as to how they complement each other in other ways, too."

He looked round the kitchen and began pulling out drawers. At last he found one in an old chest. "This will show you what I mean," he said, bringing it to the table. "Take a look at these **dovetail joints**. You can see that, by cutting the front of the drawer to fit snugly into the two sides, the craftsman has made the joints really strong, so that they don't just rely on glue when the drawer is pulled out. Husbands and wives are designed to 'interconnect' in that same close way. Their bodies interconnect, and so their lives should too."

"Hey! You're not sneaking in the suggestion that husbands go out to work while the wives stay at home are you?" Josie said, eyeing him. "What's the point of me being educated if I end up tied to the kitchen sink?"

Dad laughed at her indignant face. "I've never known you, or Mum for that matter, tied to any sink!" he commented. "Of course women should be educated and have careers, but it's no bad thing for a couple to have to rely on each other for money, however that's worked out. And you'll find that many women—and men—are actually happier in the traditional roles. After all, nurturing is written into a woman's body, isn't it?"

"Because we're the ones who become pregnant?" Josie asked.

"And carrying a child inside you for nine months is more than a bit of a distraction. But it doesn't end there, does it? After that, women are designed to feed their babies themselves, preferably for a good long time. That's nature's way of saying that it's good for both mother and child to spend a lot of time together. A man, by contrast, is bigger and stronger, and it's natural for him to go out of the home to protect and provide for his family."

"You don't know how old fashioned you are, Dad!" Josie exclaimed. "If you said that in public, I'd want to hide!"

"But you admit I've got a point," he smiled at her. "It's true all this has been educated out of young people, but nature's hard to beat. Most women only discover their priorities changing when their first child arrives—one day you'll thank me for warning you in advance."[8]

Josie considered this before saying, "You don't think I should have a professional career, then?"

"I didn't say that—it's just that it helps when you're planning your future to remember that one day you may want to dovetail your work round family commitments. You may want a job where it's easier to take time off for a bit, or work from home. Don't you believe all those stereotypes about life in the home being boring. Women I know do all sorts of things, paid and unpaid. They're particularly good at running small businesses, and doing things that involve working with people, including understanding customers.

"Men, by contrast, are obviously better at jobs which require physical strength, but they're also good at organisation and working with systems. You don't get many female air traffic controllers, for instance."

He turned the drawer round. "Let's have a look at this drawer again. You can, of course, run your married life with a minimum of interdependence, like these joints at the back here. You can see that the cabinet maker has saved himself trouble by just using glue. If the husband and wife don't actually *need* each other, don't rely on each other in their everyday lives, they're a bit like this joint, sticking together only with the 'glue' of their feelings."

"And the feelings can run a bit dry after a time?" said Josie.

"That's true," said Dad, getting up to put the drawer back, "but it's also dull. The romance which keeps love young is much more evident where the two sexes are allowed to complement each other. And we do this in all sorts of ways: for example, in the way we each think, the way we act, even spiritually. Dr Peterson told us that males and females are different even as sperm. The males get there fastest, but run out of steam and die, while the females take a bit more time to arrive, but have more staying power."

"I like it," said Josie, grinning. "I knew girls were tougher than boys!"

"Yes, you live longer in the female tract as well as in old age. So you see that males and females are different from each other, but act together, in lots of ways. By the way, you don't have to be married to live complementarity spiritually."

"But I thought spirit and body went together," exclaimed Josie.

"They do, within the family. But all of us, married or not, are either male or female, and all of us have the capacity to give ourselves to others, while remaining complete individuals in our own right. It's this give-and-take which is expressed in the fact that none of us are self-sufficient. We actually become more *fully* ourselves by giving ourselves to others."

"So," Josie said slowly, "you mean that, by looking at the body—which is either male or female, and so isn't fulfilled in itself—you can also see that spiritually we're fulfilled by giving ourselves to other people?"

Dad nodded. "But, in a spiritual relationship, you don't have to be bonded to just one other person, in the way you do when you're married. And the fact that you're not married leaves you freer to give yourself to others. For example, some people do so in a religious vocation. You'll find that, in Christianity, it's common for consecrated people to be given a title taken from family life: Father, Mother, Sister, Brother. But there are many other single people who play an invaluable role in society, and help things tick."

"I hadn't thought of single people like that," said Josie, "or of sex, for that matter."

"What I'm sure you'll begin to recognise is the tension we all feel between wanting intimacy and acceptance—the sexual side, if you like—and wanting our own space, being independent. You'll notice it especially now when you're beginning to find your way in the adult world.

"Anyway, cheer up," he said, clearing the table and picking up the apple core, "it'll be all girls tonight. What time do I come to collect you?"

Points to remember

Sexually transmitted diseases (STDs) are also known as infections (STIs) to emphasise that they can be present and passed on without showing symptoms.

Often an STD can only be detected by testing; it is wise to attend a GUM clinic after any risky sexual encounter, since infections are easier to treat if found early. Untreated, many create serious problems, including infertility.

STDs can be caught in a first encounter. They often come in multiples.

Condoms help protect against some infections, but afford no protection against many others.

HIV/AIDS is the most serious infection but it is rarely found among school children in the UK.

Noticeable early symptoms of STDs include: discomfort or pain passing urine; and pain or irritation in the genital area, swellings, ulcers and unusual discharges.

Itchiness in the genital area can come from urinary infections. Penile irritation will usually clear up by washing under the foreskin.

Bacterial infections can usually be treated with antibiotics. The symptoms of viral infections can be treated, but infections sometimes remain permanently.

Bacteria, like all other organisms, are programmed to survive; they are now adapting to resist antibiotics. The more antibiotics are used, against any illness, the less effective they become.

Young people's immune and reproductive systems only reach maturity as they approach adulthood. Adolescents are more prone to STDs than adults.

Men and women are different from each other in mind and spirit as well as in their bodies. They are designed to complement each other, in the family but also in society as a whole.

Single people are in this sense no less sexual in their behaviour than married people and, being freer, can give richly to society.

GLOSSARY

AIDS	Acquired Immune Deficiency Syndrome. A disease caused by the virus **HIV** (Human Immunodeficiency Virus). The body's immune system breaks down, making AIDS patients increasingly more vulnerable to other infections and diseases. New methods to combat HIV/AIDS are continually being sought.
Bacteria	Single-celled organisms which can live and multiply on any surface. Only some are harmful. Bacterial infections can usually be treated with antibiotics if caught in time.
Chlamydia	Bacterial infection and one of the most common STDs. Easy to cure but may remain unnoticed in both women and men until it spreads to other parts of the body. Early signs of infection include pain and discharge. May cause Pelvic Inflammatory Disease and infection in the testicles, leading to infertility.
Condom	Enclosed latex 'sleeve' which fits round the man's penis to prevent seminal fluid from touching the female organs. Has a significant failure rate in preventing pregnancy as well as against disease, especially when used by young people.
Dovetail joints	Commonly used to join sides of a drawer to the front. 'Pins' from one board fit at right-angles into 'tails' cut out of the other, often in a fan shape resembling a dove's tail.
Foreskin	Fold of skin covering the end of the penis.
Genital warts	Common infection caused by **Human Papillomaviruses (HPV)**. Genital warts may clear spontaneously, but treatment can include creams, freezing and laser. Since several types of HPV are known to cause cancer of the cervix, vaccines against HPVs have been developed.
Gonorrhoea	Bacterial infection resulting in burning at urination and discharge from the penis or vagina (women are often without symptoms). Untreated, it can lead to Pelvic Inflammatory Disease in women and painful testicles and prostate in men, reducing fertility in both. Rarely, it affects the whole body. Some strains are becoming resistant to treatment.
GUM Clinic	Genitourinary Medicine (or Sexual Health) Clinic.
Genital Herpes	Viral infection which can remain in the body without being noticed for months or years. Symptoms may be similar to those of flu, with additional stinging and sores in the genital and anal areas. Anti-viral tablets can ease discomfort, but do not cure the infection whose symptoms tend to recur.

Pre-eclampsia	Condition only known in and after pregnancy. Symptoms include raised blood pressure and presence of protein in the urine. In serious cases, it can risk the life of both mother and child.
STDs	Sexually Transmitted Diseases are infections passed on through sexual contact. Intimate behaviour, without necessarily including intercourse, is enough to pick one up. Common ones among young people in the UK are chlamydia, genital warts, herpes and gonorrhoea and several are often present together. Also known as **Sexually Transmitted Infections (STIs)**.
Syphilis	Bacterial infection which is rare but serious when it occurs. Mild initial symptoms include sores. Can be treated with antibiotics. Untreated, it can eventually affect any tissue or organ.
Virus	Strands of genetic information enclosed within a protective protein coat. Viruses are much smaller than cells, and those that are harmful invade their target cells. They then take over the cells' mechanisms to multiply and escape, destroying the target cells in the process.

Chapter 10

Forgotten Truths

"I don't get adults," Josie announced during a break in the conversation. She had become unusually quiet and it was clear that something was bothering her.

Her cousin Helen had come for lunch. It was an occasion the whole family had been looking forward to because Helen, a favourite cousin, was bringing her new husband to visit them for the first time. Josie's brother Michael had eyed Adam when the pair arrived, hoping that he might find a fellow rugby fan. However, the likelihood of watching the afternoon's match diminished as a lively lunch got underway.

Josie now pinned Dad accusingly with her eyes. "One minute we're told it's irresponsible to take drugs and the next it's irresponsible not to take them," she complained. They had been talking about cheating athletes until Josie switched the subject. She was upset because a school-mate had been put on **the Pill**.

"Oh, that's different," remarked Michael.

"Big deal it's different," said Josie, defiantly. "You boys don't have to take it. Anyway, she hates it, and the one she's been given makes her feel bloated and gives her a headache. And she's terrified her mum's going to find out."

Mum put in gently that the girl was bringing trouble on herself by having sex in the first place.

"I knew you'd say that," said Josie. "I've warned her, but it isn't just her. It's the fact that, even if you're married, women end up taking drugs when they're not even ill and men seem to get away with everything."

There was an embarrassed pause.

"May I respond?" asked Dad, ignoring Mum's glances towards their guests. Josie had said rather more than she'd meant, and was now concentrating on eating. "Fine," she mumbled.

"Well, as a man, I have to agree with you." Josie relaxed a little as he went on, "It's quite time that the whole issue of **contraception** was brought out into the open—it can't make sense to be green about everything except women's bodies. But there's something you've missed."

"What's that?" Josie asked.

"We shouldn't be giving drugs to men, either," Dad replied. "We need a whole new approach for everybody, and it'll be up to your generation to bring it about."

Helen caught Adam's eye reassuringly with a look that said, "Didn't I tell you that my uncle loves a good argument?"

Mum attempted to divert the conversation but Dad was now in his stride. "Josie has raised a very valid point." He turned to Adam and asked, "I wonder what you two would say about it?"

At this, Mum said, "Really, Paul, I don't think that this is conversation for the lunch table."

"Why not?" Dad asked. "Birth control's such an important topic. It's an interesting one, too. Just think how much social change has come about on the back of it. I bet most people don't even realise that fifty years ago it was widely frowned upon in Britain, just as it was elsewhere. D'you know, in America it used to be against the law to promote it, and in France and Ireland too?[9] Laws like that don't appear without reason. Now, by contrast, the Pill's given out free in the UK, to any woman regardless of her age or income. That's a huge change in thinking, which has been paralleled in other countries too, and I think it needs open debate."

Adam looked taken aback at all of this, and said so. He had never heard about any law against contraception. As far as he was concerned birth control was a sign of social progress, and had given women the freedom to run their own lives.

Helen warmly supported him, though she did agree with Josie that women bore the brunt. "They always used to talk about a male Pill. That seems to have dropped out of the equation now. There really should be better ways of sharing responsibility."

Michael rolled his eyes and remarked bluntly: "Well, at the end of the day it's the woman who's going to end up with the baby, so she should want to be in charge."

"I think Michael's got a point there," Dad observed, "and, much as it may be unfair, I don't think you're going to change the situation. I have another reason for saying that, which is that a woman has a remarkable capacity for putting up with things that a man just wouldn't."[10]

"Sometimes she's just forced into it," Helen said.

"That can be true," Dad agreed. "Circumstances can force her, or she can give way to pressure from the man. But, as I see it, any man who loves a woman should take an active interest in what's going on and make sure that she isn't harming herself."

"For both their sakes," Mum added. "You see, fertility is shared. If either party risks damage to their health, long term it affects them both."

"Yeah, but you've got to get real, Mum," Michael replied. "Lots of people are using contraception who aren't married or even thinking about having babies."

"That doesn't contradict what Mum and Dad are saying," protested Josie. "What they're saying is that, if a man isn't bothered about his girlfriend's health, now *or* in the future, he's not much of a lover. Isn't that right, Dad?" Josie asked.

Dad nodded. "I think that's well put. No man should shut his eyes and say that contraception is a woman's affair, even if the woman is the one who uses it."

"Well, I think it's stupid," insisted Josie, "because the woman's infertile most of the time anyway—it's the man who's always fertile."

"You haven't thought that perhaps it's women rather than men who take the Pill precisely because of their biology?" Adam asked. "You can recreate infertility in a woman, but men are always fertile so there's more to interfere with."

Dad agreed with him. "Scientists have long been looking for a good male Pill, but they haven't come up with anything that works and is acceptable. Men just won't put up with the side-effects of artificial hormones, like reduced sex drive. I believe they're now looking at totally new ideas, such as applying heat to the testicles.[11] You'll remember that sperm are very sensitive to temperature."

"Well, nobody's going to do that to me!" exclaimed Michael.

"Michael, dear," Mum interrupted quickly, "can you please keep eating. Would anyone like some more?"

Dad got up to carve. "Helen, what about you. Can I cut you some?"

Sexuality Explained

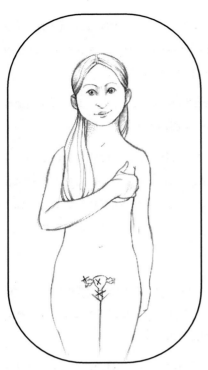

The conversation returned to lighter subjects but, after a few minutes, Dad wiped his mouth purposefully and said, "So, you'll find that most people get carried along because contraception is such an easy idea—children when you want and sex whenever you want—but that's not good enough. You've got to know what you're doing and what the consequences might be."

Helen shook her head at him. "Uncle Paul, you won't give up, will you? Well, I'll tell you about the Pill and how it works. It stops the egg ovulating."

"Among other things. It can also slow the movement of the tubes," said Mum.[12]

"Really?" asked Helen, interested. "But how does slowing the tubes stop conception? Surely the egg still reaches the uterus."

"I'll explain," said Josie, "at least, I think I can. I don't know much about the Pill, but if the tubes go slow, the embryo—that's the fertilised egg—will get stuck and then it'll just run out of food.[13] You see, the embryo carries with it all the food it needs until it implants in the uterus."

Michael furrowed his brow before saying, "I reckon conception can be stopped in three ways."

> **The three ways to stop pregnancy are to:**
>
> Stop the sperm reaching the egg
>
> Stop the egg reaching the sperm
>
> Stop the fertilised egg implanting and growing in the uterus.

Dad looked up enquiringly, but Adam got in first, saying "What do you mean, only three? There's a much wider choice than that. There are various sorts of Pill, then there are injections and gels, there are **condoms**, there are **IUDs**, oh, and **diaphragms** and other **barriers**."

"That's not what I mean," replied Michael. "What I'm saying is that, whatever method you use, there can only be three basic ways in which they all work."

"Come on," said Dad, "you tell us."

"Well, the first way's obvious. You just stop the sperm reaching the egg."

"Right," said Dad. "So you want us to imagine a big X in front of the cervix."

"And the next way is like the Pill. You stop the egg reaching the sperm."

"OK," said Dad, "so we can put an imaginary X against the ovary. The sperm has been blocked from meeting the egg and the egg from meeting the sperm. What's your other way?"

An IUD

is inserted into the uterus to prevent pregnancy. It appears to intercept the sperm, and to create a hostile environment in the uterus, which tries to expel the foreign body. The embryo is treated in the same way, and becomes unable to implant.

Early **IUDs** were made of a variety of materials: gold, ivory, ebony, even platinum studded with diamonds. Pebbles, wool and glass have also been used. Most modern **IUDs** are made of plastic and/or copper (discovered in the 1960s to have exceptional antifertility properties) and sometimes also release hormones.

Shape and size are important: too small, and they are less effective and may be expelled; too big, and they may perforate the uterine walls. The withdrawal thread can become a route for bacteria from the vagina to penetrate the cervical gate, causing infections. Other side-effects include abdominal pain, heavy bleeding and ectopic pregnancy.

"The third one is stopping the fertilised egg from implanting."

"So that means a big X inside the uterus. Does anyone have any other ideas?" Dad asked.

Adam spoke up again: "What about IUDs? They sit in the uterus, but their action is to stop the sperm from reaching the tubes."

"So that's really method 1, even though it happens in the uterus," said Dad. "Actually, scientists still argue about IUDs. They appear to block sperm sometimes, but that's not all they do. They also prevent implantation—which is method 3."[14]

Mum turned to Michael, saying, "I'm impressed. You know a lot to be able to work all that out."

"No he doesn't," Josie countered, "that last one isn't stopping **conception**. It's stopping **implantation**, which happens about a week after the baby's been conceived."

"You know that, Josie," Mum explained, "because you've seen the biology. Many people have little idea of how the body actually works. All they know is that, if you take certain precautions, intercourse is less likely to result in pregnancy, and they'll call anything that does that contraception."

"But look," said Adam, "it's not surprising people are still working things out—this is all new science and the world is changing fast. It'll settle down some day and people will learn more."

"I don't think so," said Mum, "it's a forgotten truth that contraception's always been around. The idea of IUDs goes way back, to the Ancient Greeks even, and it's long been known that they can bring on an early period."

Josie was still musing over Michael's three Xs. She reckoned that there was a fourth way of avoiding a pregnancy which he'd left out.

"Surely, if the woman's only fertile for a few days each month, all you have to do if you don't want a baby is to know when those days are and avoid them?" she said out loud.

"OK," said Dad, turning to her, "so where does the X go?"

Michael thought about this before saying, "There isn't one."

"Except on the calendar," Adam muttered.

Dad raised an eyebrow at him. "So you don't need contraception to avoid pregnancy. All you need is some planning and a bit of discipline, as Adam points out, but, provided you really know when the fertile days are, and understanding the woman's mucus cycle gives the key to that, you can plan pregnancies without harming anyone's body."

"Josie's raises a really important point in another way," added Mum. "You see, giving life to another human being is something very precious and touches the depths of who we are as people. It calls for respect. Fertility shouldn't be treated as though it's a handicap."

"Which is why artificial birth control has always been controversial," said Dad, "interfering as it does with intercourse or maiming fertility in a healthy body."

"Do you have to use such emotive language, Uncle Paul?" exclaimed Helen.

> **Sterilisation**
>
>
>
> Both men and women can be made permanently infertile by cutting and tying the tubes which carry the man's sperm (called **vasectomy**) or the woman's eggs (**tubal ligation**).
>
> This is the most popular form of birth control in older couples in the West. It is also widely used in populous developing countries, sometimes under programmes of compulsion.

"Well, looking at it through the ages, I would call contraception pretty brutal. Everything we have today has an ancient forerunner, you know. Take IUDs for starters. Mum's said that they go way back. In fact there are descriptions of them dating to the time of **Hippocrates** in the fourth century BC. All sorts of bits and bobs were used to block the woman's uterus—and the cervix wasn't hygienically dilated to make their passage more comfortable." He raised his eyebrow adding: "Believe it or not, IUDs were also inserted into camels, to stop them conceiving on long treks through the desert."

"Yes, Uncle Paul, but that's not today," insisted Helen.

"True enough. Today's versions are usually made of plastic or copper, and they're altogether better designed. But look at the possible side-effects: heavy bleeding and severe pain are common, and then there's perforated uterus, **ectopic pregnancy**, **Pelvic Inflammatory Disease**,[15] to name just a few, let alone the complication of when a baby's conceived despite everything. Do you remove the IUD or do you let it be? I just wouldn't want your mum to risk all of that."

"I agree that IUDs are a bit barbaric," Adam commented. "I wouldn't want Helen to use them either."

Helen was amazed. She looked across the table at him. Whenever she had tried to raise the subject of contraception in the past, Adam had dodged giving an opinion, saying that it was 'a woman's business'. If she teased him about it, he would look at her lovingly and tell her that it was her body, and he would support her in whatever she did.

Recently she had been getting bouts of depression which were so unlike her. She hadn't yet said anything to Adam, but she had a feeling they could be associated with the Pill. OK, they weren't that big a deal, but the other day she'd idly Googled the IUD and been quite interested. She'd looked at the comments section. Some users were pleased, some weren't, most had experienced pain and bleeding at some point, especially when they were inserted. But the bottom line for these women was always that IUDs were better than the side-effects of taking hormones.

She switched her mind back to the table. Her uncle was saying. "And guess what comes with multiple partners?"

> ### Spermicides
>
>
>
> are chemical **'sperm killers'**. They, with barrier methods, are some of the oldest and simplest forms of birth control, and are often used together.
>
> A curious mixture of substances have been used to immobilize and destroy sperm. In the nineteenth century BC, the Egyptians were using mixed honey, natron and crocodile dung. During the Middle Ages, people tried rock salt and alum, and in the early twentieth century Marie Stopes was recommending vinegar, carbolic soap and quinine solution.
>
> Surface-active agents, which breakdown the sperm membrane, were developed in the 1950s and are now the commonest active ingredients in UK products. Spermicides come in the form of creams, foams, gels, films and pessaries, and can cause irritation and occasional infection.

"STDs? Come on, Dad. You'll be telling us next that they used condoms years ago too," said Michael, with a grin.

"Spot on!" said Dad. "How did you know that? The Romans got there before us, thinking that a waterproof barrier would prevent germs spreading. They didn't have latex, so what d'you think they used?" He paused. "I'll give you a clue—it's a very fine skin, and it comes from animals."

"The intestines, I suppose," said Adam.

"That's yuk!" exclaimed Josie.

"But it's true. We know that animal bladders were used by the Romans to protect from STDs," Dad told them.

"Weren't they used for the first rugby balls, too?" Michael asked.

"I believe they were, acting like balloons inside a leather casing—and I'm glad to say that they've been replaced by rubber, too. You can see how materials develop but the ideas remain basically the same. There's evidence from various countries over many centuries of condom practice, used against disease and pregnancy."

He added softly, "Other experiments are too horrible even to mention, except to say that every sort of obstruction and chemical has been tried. And, from the very beginning, methods were used interchangeably to cause early abortion as well as contraception."

"With little thought for the woman," Mum responded smoothly. Then she added in a brighter voice, "But, Paul, we didn't say that for many centuries all this affected very few people. Social mores underwent huge changes with the coming of Christianity. Contraception didn't disappear, but it went underground. It's difficult to conjure up now just how alien the idea was to most ordinary people for hundreds of years."

"Well, things were pretty primitive then, as you were saying," said Helen. "Now contraception's just part of life, and you're going to have to accept it. I agree it's not a good idea when you're at school, but, once you're grown-up, I can't see the problem."

Condoms and Caps

Since time immemorial, attempts have been made to intercept sperm in the vagina. Barriers have been concocted from many materials, including animal gut, sponge, cloth and paper. Modern latex is derived from vulcanized rubber, first discovered in the mid-nineteenth century.

The most common barriers are: **condoms** (usually used by the male) and **cervical caps/diaphragms** (used by the female). Barriers have fewer side-effects than other contraceptives but can be uncomfortable and cause skin irritation. They are disliked for interfering with the sexual act.

The failure rate of **condoms** and **caps** varies widely with frequency of intercourse, fertility and carefulness of use. Among young people, the pregnancy rate is about 15% a year.

STDs can pass at any time of the month. Viruses are smaller than sperm (the **AIDS** virus is about 25 times smaller than the sperm cell's head) and can penetrate natural holes occurring in the latex. **Condoms** appear to reduce the risk of some **STDs** but are far from giving full protection.

Condoms are also sensitive to environment and deteriorate easily if, e.g., left in a warm pocket or cold car. Theoretical effectiveness cannot be relied upon.

"You can't turn the clock back," Adam joined in. "Everybody uses contraception. Think about the Pill. Just about every woman must've used it at some point."

"Hey, Dad," Michael said with a grin. "What about the Pill? What sort of Pill did the Ancient Greeks have?"

"Well, it didn't come in packets," Dad returned. "But you'd be amazed what herbs and potions the Greeks and Romans used. One was called **silphium**. It only grew in Libya and was apparently so effective that, by the second century AD, it had been picked to extinction."[16]

"So you're saying that some herb was as good as all the science put into modern pills?" asked Michael.

"I've no idea," said Dad. "Saying you won't have a baby isn't the same as saying you'll enjoy all the other consequences."

"Just like the Pill," said Mum.

"What do you mean, just like the Pill?" said Helen defensively. "Millions of women are on the Pill, and you can't say it doesn't work."

"What does it actually do?" asked Josie.

"The idea's simple enough," Mum replied. "When a woman's pregnant, high levels of progesterone stop any more eggs being released—nature's way of preventing births overlapping. So what the Pill is designed to do is to fool the body into behaving as though it's pregnant by giving it extra hormones. But you need a very large dose to override the normal cycle. Finding a cheap source of progesterone which could be taken in that quantity in pill form created a problem for scientists. In the end,

The Pill

was developed in the USA and launched in 1960 as the 'magic' contraceptive which would do away with the messiness of IUDs, barriers and spermicides. Its invisibility did much to make contraception more acceptable.

It works by using hormones to give false messages to the ovaries, endometrium and cervix so that they stop their normal functions. Dr John Rock, who helped create the Pill, realised that women would feel uncomfortable without regular bleeds so he designed the dosage to allow for mock 'periods'.

Experience shows that it is easier to prevent the embryo implanting than it is to stop the egg's release. Ovulation breaks through with all hormonal contraceptives, which then rely for their effect on the changes they make to the cervix or in the tubes and endometrium. Break-through ovulation is especially associated with progestagen-only methods, such as the Minipill and long-acting injectables.

Because the Pill acts at cell level, many side-effects are associated with it. These include serious conditions such as cancer, deep-vein thrombosis, strokes and heart attack. More common side-effects include nausea, bleeding, headache, depression, loss of interest in sex and weight-gain. Exactly what is attributed to the Pill is disputed.

The Pill is popular for its convenience and potential effectiveness. However, method-effectiveness failure rates of less than 1% pregnancy per year become more like 8% in actual general use and 11% among younger users.

---

they've found it easier to work with artificial testosterone, and use it to mimic progesterone. These artificial hormones became known as **progestagens**, but in the dosages used they're many times more powerful than natural progesterone."[17]

Josie made a face at this. "So that's how they stop the egg meeting the sperm?" she asked.

"Yes, and no," said Mum. "Progestagens on their own led to too much bleeding, so they decided to add in oestrogen as well—it was initially left out because it was known to stimulate rapid cell growth, which causes cancer."

"And what happens in the rest of the body?" Michael asked. "I thought those hormones controlled all sorts of things."

"Good question," Mum replied, "because the hormonal balance in a woman is very delicate."

"The Pill is certainly no 'on/off' switch," Dad agreed. "Just pick up a textbook designed for practitioners and run your eye down the index and you'll have a good idea of its drawbacks."[18]

"There's one important way in which the Pill works differently from the way it was first designed," said Mum. "The doctor behind it was a religious man and he wanted a drug which

> **The Cervix Ages...**
>
>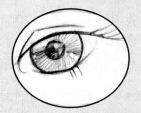
>
> ... in the normal course of a woman's life, so that it produces less of the fertile mucus necessary for conception. Pregnancy rejuvenates it, making the cervix of a 33-year-old resemble that of a 20-year-old. This explains why older women who have had children are more fertile than those of a corresponding age who have not.
>
> **The Pill's** action on the cervix creates artificial quantities of barrier-type mucus, the second of its contraceptive effects. The crypts which produce this mucus grow to an abnormal size while those which produce fertile mucus shrivel from disuse. Thus the cervix ages prematurely and the cervical canal narrows. Restoration after stopping the Pill takes time; some crypts may be permanently damaged.
>
> The distortion of the cervix reduces a woman's fertility. It also makes it more difficult to chart her mucus symptom as an aid to planning or avoiding pregnancy.
>
> *"The cervix is a precision organ as complex as the eye."*
>
> (Dr Erik Odeblad)

he thought would be acceptable to the Catholic Church.[19] So it was really important to him that the Pill suppressed ovulation rather than made for an early abortion. But in practice *all* versions of the Pill sometimes slip up. Then they rely on other effects. This includes changing the mucus, making it into a barrier,[20] but it also includes changing the endometrium and the movement of the tubes, so that the embryo can't implant."

"Hmm. So the Pill uses all three Xs on our imaginary diagram?" Josie asked.

"You've got it," Dad replied.

"Which means that it also acts by inducing early abortions?" she persisted.

"Right again. That's how the **'morning after' pill** works too—it's an extra strong version of the contraceptive pill," Dad explained.

"And I thought that contraception was meant to stop abortion!" Josie exclaimed.

"Sadly, it never has," Mum replied. "Even the Greeks and Romans were aware of that. You'll also find that the people who promote contraception are usually the same people who promote abortion."

"I started this meal feeling depressed," Josie exclaimed. "Now I feel really depressed!"

"It would be depressing if contraception was necessary," said Dad. "But actually, it isn't."

By now Adam was looking increasingly irritated. He glanced towards Helen's flushed face, and said heatedly, "Look. Everybody would love to have children. But sometimes it's just not practical. You have debts round your neck, a house to buy, a job to keep. And can you imagine what this country would be like if everybody suddenly started breeding like rabbits?"

> ### Nature's Way
>
>
>
> It is man's original ecology to reproduce every 3-5 years and have about 5 or 6 children. The children are spaced by breastfeeding, which suppresses ovulation. This saves the mother 'feeding' two babies at once. Infertility lasts longest when the mother gives suck frequently, sleeps with the child, is herself physically active, and has a good but not over-generous diet.
>
> Fertility has risen with the spread of bottle feeding. However, breastfeeding continues to be an important means of spacing children, especially in poorer countries. In richer countries, a woman's life-style tends to mean that fertility is suppressed for less time, even when she is breastfeeding.

"But they wouldn't," Dad calmly replied. "It's possible to manage your fertility without contraception. In fact, better than with it. After all, contraception is full of false promise."

Mum started gathering in the plates and suggested a complete change of topic. But Helen said, "What do you mean, Uncle Paul?" And, looking towards Adam, she said, "We may not like it, but there's a lot of truth in what they've said so far. You can't argue against something by just shutting it out."

"I'll tell you what I mean," her uncle replied, smiling warmly at her. "People use contraception thinking it's foolproof. They really believe that they won't get pregnant, and then, hey presto! along comes a baby and they don't know what to do.

"Statistics are just statistics on a page, but an individual girl's playing with her life. Look at the Pill, for instance. The overall effectiveness rate might be 99.6% and that's what's being sold to women, and to schoolgirls too. But look a bit closer, and you'll find that the 99.6% was drawn from well-motivated older women, who are naturally less fertile than a youngster. The true figure for schoolgirls could be more like 90%, which is being generous.[21] Now that sounds good, doesn't it, till you turn it round. It means that 10%, or one out of ten girls, on the Pill is likely to become pregnant in the first year. Now obviously the longer you're exposed to that possibility, the more likely it is to happen. If you extend the 10% annual rate over five years, you get two out of five becoming pregnant. That means approaching half of the girls who start the Pill aged 15 becoming pregnant at some point by the age of 20—that's not a risk I'd care to take."

Helen thought about this. She thought back to her own school days and remembered her friend Lydia. There were others as well. One had complained, "Of course, I couldn't take the Pill until the start of my next menstrual cycle, and I didn't dare tell Jack I wasn't already on it."

Her own mother had always been at home for her when she came back from school—much more effective than any contraceptive! It was later that she'd started having sex, when she and Adam were at college together. She remembered one occasion when her period had

> # Natural Methods of Family Planning (NFP)
>
>
>
> teach the couple to understand the woman's pattern of fertility, so that they can time intercourse to achieve or avoid pregnancy.
>
> The 'Rhythm' or 'Calendar' method was first described in the 1930s, and used the woman's pattern of previous menstrual cycles to predict when ovulation was likely to fall. It worked best for those with regular cycles—and was better than many other methods around at the time!
>
> However, it failed those with irregular cycles or who had recently given birth. What was needed was a method based on the present cycle.
>
> In 1964, Drs John and Lyn Billings made a discovery which revolutionised NFP. They found that mucus, which is easily observable mid-cycle, appears **before** ovulation and can be used to predict the release of the egg. With other close collaborators, they also discovered the key role that mucus plays in fertility.
>
> The Billings Ovulation Method (BOM) is now widely taught in many countries of the world. It has been adapted and added to by others, notably in the Symptothermal Method and Creighton Model of FertilityCare.
>
> All these three methods can be used to achieve or avoid pregnancy. They are recognised by the World Health Organisation as having effectiveness rates for avoiding it as high as that of the Pill.

been late. He'd asked another girl to dance and Helen had been so cross she'd stopped the Pill immediately. And then he'd come back, and he'd been so sweet, and he'd even proposed to her and ... well, she'd been lucky.

"But what about natural methods of family planning?" she asked now. "I've always been told that they don't work."

"Then you've probably been given rates for the **rhythm method**!" laughed Mum. "The best natural methods are as good as the Pill—99% and more.[22] How they work in practice depends on commitment. But I can assure you that there wouldn't be 40,000 NFP teachers working in One-Child Policy China if it didn't work."

"There's another way in which NFP is successful, too," said Dad. "You see, it's completely honest. You know when you're fertile and when you're not, and it's up to you to take responsibility. And when you do decide you would like to have a child, you know when to aim for it. Now, think back to the Xs on our imaginary chart. Contraception is also called Family Planning. But how in practice does it help you to have a baby?"

"It doesn't," said Helen simply.

"It can't help, but it can damage. It's common sense that you can't blast a delicately balanced system with pills and devices without expecting some fall-out. And the damage is there to be seen. Just look at the number of people who find themselves infertile after messing themselves up with contraceptives and STDs. Then they're desperate for babies.

"But I'll tell you the biggest cheat of all. Contraception says you can have sex when you want it and it'll make you happy. But it doesn't do that either. The most remarkable thing about NFP is that it promotes unity within the family." He smiled across at Mum. "The divorce rate for couples using it is something like 1%–4%—beat that in the modern world."

* * * * * * * * *

After Helen and Adam had left, Dad shook his head as he remarked, "Isn't it an amazing quirk of human nature. Contraception is regarded as a freedom, and in practice nobody likes using it." Then he added, "What people are really after, of course, is a mirage—the pleasure of sex without strings attached."

"And what they miss," said Mum, "is that the most rewarding sexual experience comes from having no barriers to commitment."

Michael went with Dad to collect his little sister from a friend's house. Josie volunteered to stay behind to help Mum in the kitchen. They decided to have a cup of tea before tackling the washing-up.

"Mum," Josie asked. "Why's it necessary to spend most of your married life avoiding having a baby? I mean, it's going against nature, and for what? When I grow up, I want to have a family like Amy's. It's big, but it's one of the happiest I know."

Mum smiled at her. "It can be practical to avoid having a child, but that's not where the fun is. The joy is not in avoiding, but in conceiving a child. It's such an amazing experience it changes you for life."

Points to remember

Man was originally designed to reproduce every 3-5 years and have about 5 or 6 children. Breast-feeding suppresses fertility, but varies in effectiveness according to the mother's lifestyle. Breast-feeding continues to be an important means of spacing children in many societies.

There are three basic ways to stop pregnancy artificially: stopping the sperm from reaching the egg; stopping the egg from reaching the sperm; and stopping the fertilised egg from implanting in the uterus.

Contraception is an ancient idea. Most modern methods have earlier antecedents.

Methods of contraception which prevent the fertilised egg from implanting in the uterus act as abortifacients. These have been included among contraceptive methods from the earliest times.

Pregnancy can also be avoided by planning intercourse so as not to use days known to be potentially fertile. Modern methods of natural family planning are based upon understanding the woman's mucus cycle.

Both men and women can be made permanently infertile by cutting and tying the tubes which carry their sperm (vasectomy) or eggs (tubal ligation). Sterilisation is the most popular form of contraception among older couples in the west, and is widely used in populous developing countries, sometimes under compulsion.

Intrauterine devices (IUDs) prevent pregnancy by intercepting sperm in the uterus, and by creating a hostile environment there so that embryos are prevented from implanting.

Barriers include condoms, cervical caps and diaphragms. Their effectiveness varies widely with frequency of intercourse, fertility and care of use. Condoms reduce the risk of some STDs but give little protection against others. Barriers are usually used with spermicides to increase their effectiveness.

Hormonal methods of contraception are designed to fool the woman's body into behaving as though it is pregnant. Artificial progestagens, with or without oestrogen, are delivered to the woman's body in various ways, especially in pills and injections, and act on the ovaries, endometrium and cervix. Their effectiveness in practice, especially among young people, is much less than their potential.

The cervix ages prematurely with use of the Pill, sometimes permanently.

Professionals in the field often lament that young people dislike contraception and mistrust its operation. This, and the irregular nature of youthful relationships, lead to a lack of rigour in its use and contribute to high pregnancy rates.

GLOSSARY

AIDS — Acquired Immune Deficiency Syndrome. A disease caused by the virus **HIV** (Human Immunodeficiency Virus). The body's immune system breaks down in AIDS patients, making them increasingly more vulnerable to other infections and diseases. New methods to combat HIV/AIDS are continually being sought.

Barriers — A collective term for the devices which physically hold back the seminal fluid from the cervix. Modern ones are made from latex, the most common being condoms, diaphragms and cervical caps.

Conception — The moment when a new human life begins. Historically arguments differed on when conception occurred because the reproductive process was insufficiently understood. Today the science is agreed upon, but there is confusion as to whether or not 'the process of conception' is complete at fertilisation or whether it is only complete at implantation. The confusion allows methods which stop implantation of the fertilised egg to be called contraception; it was deliberately introduced in 1965 to make the Pill more palatable.

Condom — Enclosed latex 'sleeve' which fits round the man's penis to prevent seminal fluid from touching the female organs.

Contraception — Collective term for all artificial methods of intercepting pregnancy before implantation of the egg.

Diaphragm — Latex barrier placed inside the woman's vagina before intercourse. **Cervical caps** are similar but smaller.

Ectopic pregnancy — Where the embryo implants outside the uterus (usually in the tubes). The condition is dangerous for the mother, and the embryo rarely survives.

Hippocrates — Greek physician who is considered to be the father of medicine (c.460–c.377 BC).

Implantation — Process by which the fertilised egg takes root in the wall of the uterus and begins to feed from the mother. This usually happens 6-9 days after fertilisation.

Intrauterine device — Object, usually of plastic and/or copper, positioned inside the uterus to prevent pregnancy. It acts by intercepting sperm and creating an environment hostile to embryos, which thus fail to implant. Modern devices sometimes also release hormones. Known as an IUD.

Morning-After Pill	Extra strong version of the contraceptive pill, usually taken in two doses within 72 hours of intercourse. Its principal action is to prevent implantation of the embryo. Also known as the MAP and Emergency Contraception.
Pelvic Inflammatory Disease	Inflammation of the uterus, tubes and/or ovaries, leading to scarring and adhesions. Important cause of infertility. Known as **PID**.
Rhythm method	Method of Natural Family Planning developed in the 1930s, which gives rules on when a woman can count herself infertile based on the history of her previous cycles. Also known as the **Calendar Method**.
Silphium	Herb grown in Libya and used in Ancient Times as a contraceptive potion. Picked to extinction by second century AD.
Spermicides	From the Latin for sperm-killer. Chemical substance used on its own, or with barrier contraceptives, to kill sperm within the vagina.
Sterilisation	Cutting and tying the tubes which in a man carry the sperm (vasectomy) and in a woman carry the eggs (tubal ligation) to suppress fertility permanently.
The Pill	Drug composed of artificial hormones (progestagen and sometimes oestrogen) in varying doses which act on the ovaries, tubes, uterus and cervix to inhibit ovulation, block access of sperm, and prevent implantation.

Notes

1. For an excellent and readable overview of the sexual chemistry of the brain, see Joe S. McIlhaney and Freda McKissic Bush, *Hooked: new science on how casual sex is affecting our children* (Northfield Publishing: 2008).

2. Patrick Fagan, when he was Senior Fellow at the Heritage Foundation in the US, found that women who have only ever had one sexual partner have an 80% chance of a successful marriage. Add in one extra partner and the figure drops to 54%, a second and it drops to 44%, with further drops thereafter. (Taken from his address 'The Dignity of the Child from Conception and its Right to Life, Home and Family', given at the World Congress of Families in Warsaw, 12 May 2007.)

3. Conception has always been understood to mean the start of new life. In 1965, the American College of Obstetricians and Gynaecologists decided to change its definition, tying conception to implantation rather than fertilisation. This was for social rather than scientific reasons, to ease the acceptability of contraceptive methods which prevent implantation, including the newly introduced Pill. The new terminology has remained controversial and has never been fully accepted. It makes the medical dating of pregnancy to the time of last period still further out of true and calls in question the concept of ectopic pregnancy.

4. "Government and survey data overwhelmingly document that married-parent households work, earn, and save at significantly higher rates than other family households as well as pay the lion's share of all income taxes collected by the government. They also contribute to charity and volunteer at significantly higher rates, even when controlling for income, than do single or divorced households, leading Arthur Brooks of the American Enterprise Institute to write that 'single parenthood is a disaster for charity'… Data from an earlier wave of the survey reveal the disparities of household income among a greater range of household types with children under 18. For 2001: intact, married families had a median income of $54,000; stepfamilies, $50,000; cohabitants, $30,000; divorced-single parents, $23,000; separated-single parents, $20,000; widow parents, $9,100; never-married single parents, $9,400." Extracted from "The Family GDP: How Marriage and Fertility Drive the Economy" by Patrick Fagan in *The Family in America Journal of Public Policy* (Spring 2010); see http://www.familyinamerica.org/index.php?doc_id=9&cat_id=7.

5. For a helpful study of same-sex attraction, see the article by Dr Richard Fitzgibbons MD entitled "Same-Sex Attractions in Youth and their Right to Informed Consent" on the Child Healing: Strengthening Families website at http://www.childhealing.com/articles/ssayouth-if-imh.php.

6. In 2000, a number of US federal agencies sponsored a major international workshop to draw together existing expertise and published research in a definitive study of the effectiveness of condoms in preventing STDs. A detailed examination was made of eight common infections. The twenty-eight experts responsible for drawing up the workshop's report agreed that consistent condom use decreased the risk of HIV transmission in both men and women by approximately 85%, and that correct and consistent condom use could reduce the risk of gonococcal transmission for men. With regard to all the others, chlamydia, trichomoniasis, genital herpes, syphilis, chancroid and gonococcal transmission to women, there was insufficient evidence to draw definite conclusions of their effectiveness. For HPV there was no evidence that they helped at all: "In spite of all the talk today about condoms and safe sex, it could not be proven that condoms offered any protection in 7 of the 8 sexually transmitted diseases reviewed. The panel reviewed the medical

literature of the past couple of decades, and was unable to prove that condoms work for herpes, human papilloma virus, trichomonas, chlamydia, cancroids or syphilis. The first three account for an estimated 12-20 million infections per year." See *Report on Condom Effectiveness* (Washington, DC: National Institute of Health, 2001).

7 This phenomenon is known as risk compensation behaviour, and is seen in a wide range of activities. A 2007 article in the Lancet (Shelton, James D. (2007-12-01). "Ten myths and one truth about generalised HIV epidemics." *The Lancet* 370 (9602): 1809–1811) suggested that "condoms seem to foster disinhibition, in which people engage in risky sex either with condoms or with the intention of using condoms".

8 Catherine Hakim's ground-breaking research *Models of the Family in Modern Societies* (Aldershot: Ashgate, 2003), revealed that women's lifestyle preferences are different from men's. Dividing women between those who are work centred, home centred and adaptive between the two, she found that by far the largest number (69%) are adaptive. This choice is intrinsic and has virtually no connection with political and religious values. More recently, the report *What Matters to Mothers in Europe*, presented by the Mouvement Mondial des Mères Europe (World Movement of Mothers Europe) to the European Parliament in May 2011, looked specifically at the desires of mothers. Based on 11,000 responses from across Europe, a unifying complaint was that nobody had prepared them for the change of priorities motherhood gave them. Three out of five wanted part-time work while they had children at home with 83% agreeing that teenagers need mothers around. There was strong feeling that caring for the family is not given the kudos it deserves and that mothers should be given more leeway to decide their own work/family life balance. http://www.mmmeurope.org/sites/mmmeurope.org/files/documents/MMM%20BROCHURE%20What%20Matters%20to%20Mothers%20in%20Europe.pdf.

9 The ban on the distribution of contraceptive devices in the USA was first introduced with the Comstock Act of 1873. It took various shapes in the intervening years before being finally overturned by the US Supreme Court for married couples in 1965 (*Griswold v. Connecticut*) and the unmarried in 1972 (*Eisenstadt v. Baird*). Contraception was banned in France between 1920 and 1967, and in Ireland between 1935 and 1980.

10 The first combined pill (using progestagens and oestrogens) was tried out in the 1950s on both men and women. It successfully stopped both sperm production and ovulation. However, the trial on men was halted after one candidate displayed shrunken testicles; later on, three of the women died but by then millions of women were already on the Pill. See Dr Ellen Grant, *The Bitter Pill* (Corgi, 1986), p. 19.

11 For reference, e.g. http://www.malecontraceptives.org/methods/heat_biology.php.

12 This effect is particularly associated with progestagen-only contraceptives, such as the Mini Pill, long-acting injections (e.g. DepoProvera), and subdermal implants.

13 The embryo may also implant in the tube, causing an ectopic pregnancy. Such pregnancies either abort spontaneously or have to be removed, with or without permanent damage to the tube.

14 World Health Organization, "Long-term safety and effectiveness of copper-releasing intrauterine devices: a case-study" (2008); John Guillebaud, *Contraception: your questions answered* (3rd edition, 1999).

15 Pelvic Inflammatory Disease (PID) is a serious condition which often leads to permanent sterility. In their chapter on IUDs, Howard J. Tatum and Elizabeth B. Connell advise women who may wish to be bear children at a later date against using this method of contraception. See *Contraception, Science and Practice*, pp. 165–166.

16 David W. Tschanz, *Herbal Contraception in Ancient Times* at www.islamonline.net.
17 See Ellen Grant, *The Bitter Pill* (Corgi, 1985). Dr Grant worked on the Pill's development for ten years when it first came to England in the 1960s.
18 See, for example, Marcus Filshie and John Guillebaud (eds), *Contraception: Science and Practice* (London: Butterworths, 1989). It is still regarded as a landmark textbook and informed the Royal College of Obstetricians and Gynaecologists' *Report on Unplanned Pregnancy* (September, 1991) on which UK government policy in favour of promoting contraception among the young was based. The chapter on the Pill, by Michael D. G. Gillmer, Consultant Obstetrician and Gynaecologist at the John Radcliffe Hospital, Oxford, begins as follows: "It is now 25 years since oral contraceptives became available for general use, and during that time our state of knowledge about how they work and what side effects they may have has progressed from a state of profound ignorance to one of relative ignorance." It is also worth remembering that trials often compare the side-effects of Pill-users with those e.g. of IUD-users rather than with controls who have never used contraception. Results can also be distorted by women dropping out when they experience early side-effects.
19 Dr John Rock was a Catholic obstetrician who worked with Dr Gregory Pincus and Dr Min Chueh Chang to create the Pill. Their work was funded by Margaret Sanger, the well-known birth control activist. Dr Rock was bitterly disappointed when Pope Paul VI published *Humanae vitae*, which reiterated the teaching that contraception of all kinds is illicit. He never went to church again.
20 Much of what is known about the action of the cervix is down to the pioneering work of Professor Erik Odeblad.
21 John Guillebaud, *Contraception: Your Questions Answered* (Churchill Livingstone), which is regularly updated with new editions.
22 See http://www.wordiq.com/definition/Symptothermal_method#Statistics.

Whole page diagrams

Illustrated by Jessie Gillick

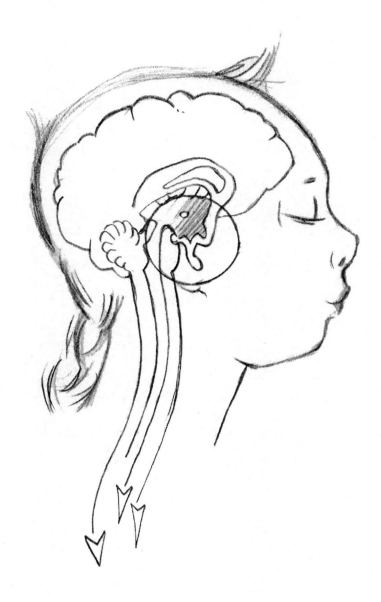

The pituitary gland in the brain controls the reproductive organs in men and women

Female reproductive organs

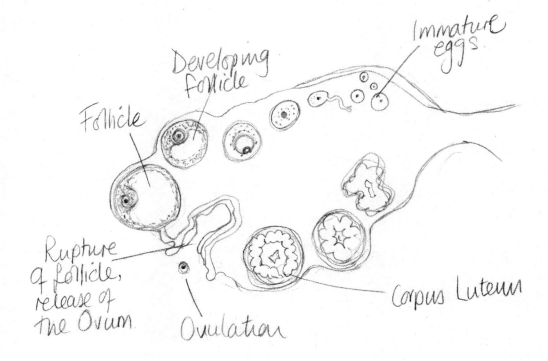

Cycle of events inside the ovary

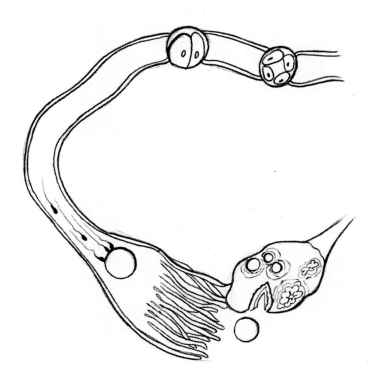

Ovulation

The fimbria pick up the newly released egg

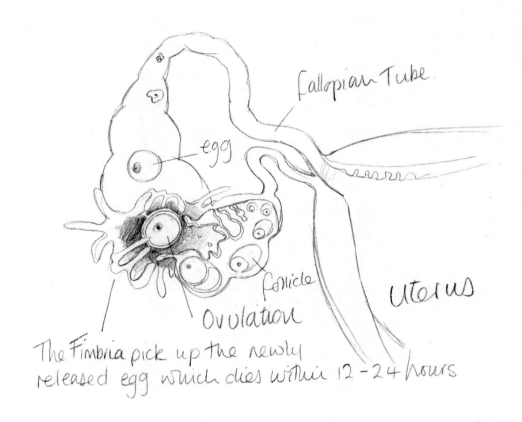

Ovulation and death of the egg

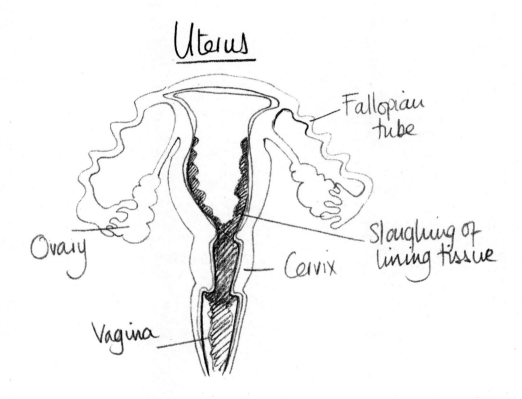

Menstrual bleeding, or period

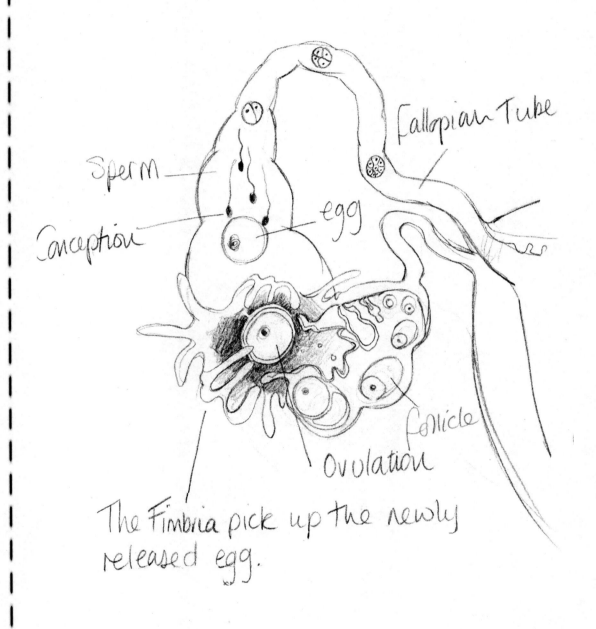

Ovulation and conception

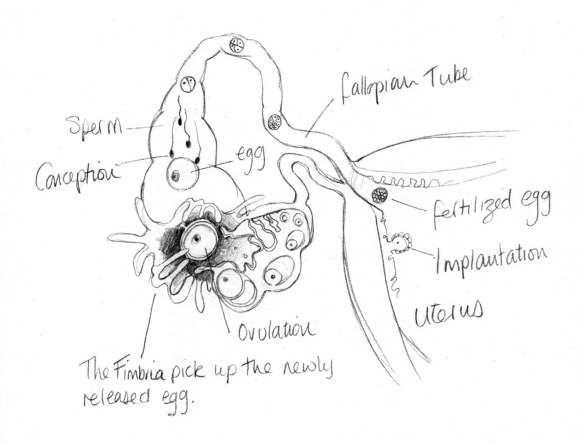

Implantation of the embryo

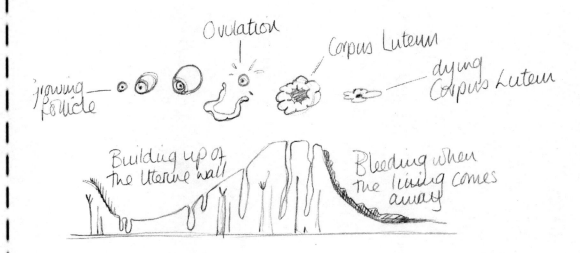

Cycle of events in the uterus with ovulation cycle above

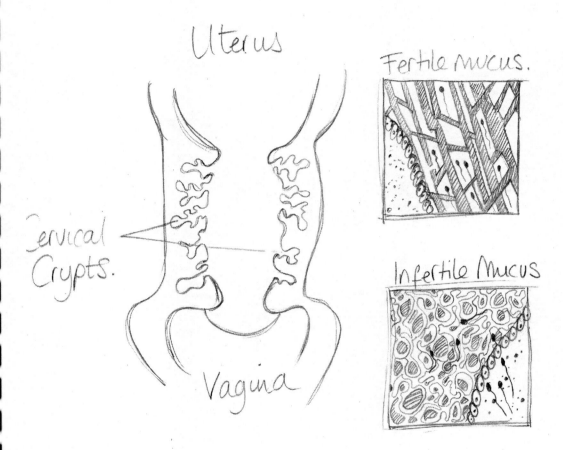

The cervix and its mucus

Fertile cervical mucus under a microscope

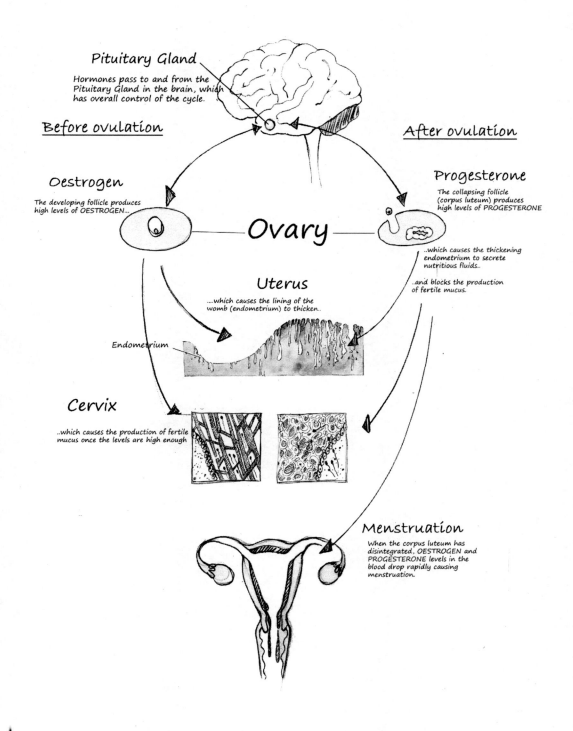

The full menstrual cycle

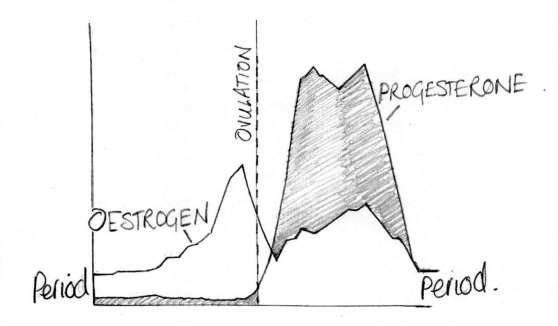

Female hormone cycle

Male reproductive organs

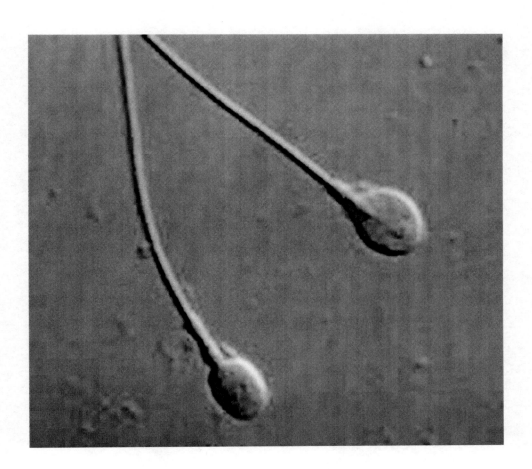

Two sperm under a microscope, showing their heads and tails

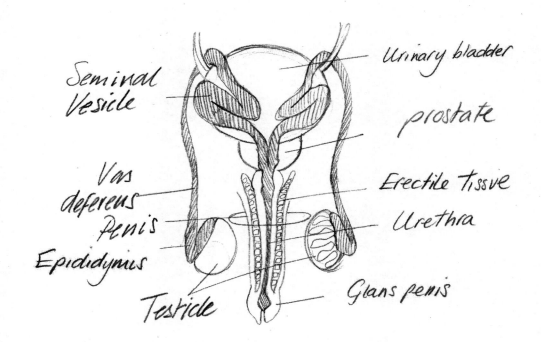

Sperm transport system

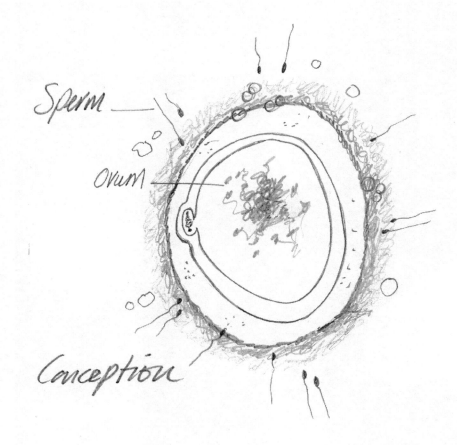

Conception

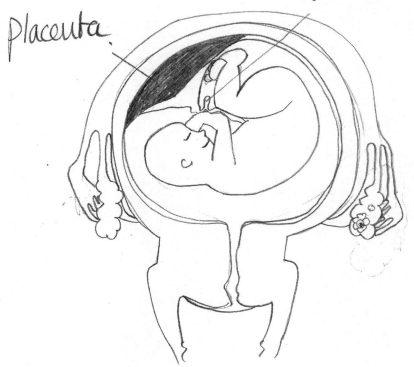

Baby at 16 weeks gestation

Lightning Source UK Ltd.
Milton Keynes UK
UKOW03f2059280415

250515UK00004B/13/P